WITH COMPUTER GRAPHICS

# Graphics & Geometry 1

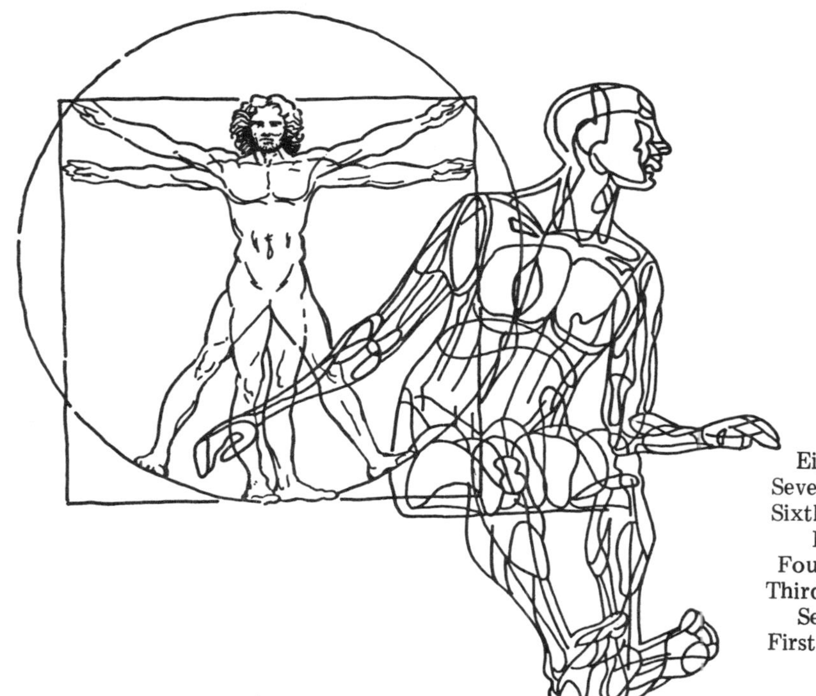

James H. Earle
Samuel M. Cleland
Lawrence E. Stark
Paul M. Mason
North B. Bardell, Jr.
George J. Raczkowski
J. Timothy Coppinger

TEXAS A&M UNIVERSITY

Eighth Printing, July 1986 (2.5W04.2)
Seventh Printing, January 1986 (2d55.2)
Sixth Printing, September 1983 (5c55.2)
Fifth Printing, June 1981 (5.3B36.2)
Fourth Printing, October 1979 (5h65.2)
Third Printing, September 1977 (5P78.1)
Second Printing, March 1976 (6v06.1)
First Printing, January 1975 (10.1E15.1)

Copyrighted by
Creative Publishing Co., 1975, 1986
Second Edtion

ISBN 0-932702-79-1

Creative Publishing Company
Box 9292, PH. 409-775-6047
College Station, Texas 77840

# To the student

GENERAL: This books is designed to present the fundamental principles that will enable the engineering and technology student to think and communicate graphically. Design problems are included to stimulate creativity and the exercise of imagination so often required of the engineer.

COMPUTER GRAPHICS PROBLEMS: On the backs of the sheets are problems covering the same principles on the fronts that are to be solved by computer graphics. This double approach will enable you to solve problems by either computer graphics or traditional graphics, or both. Additional instructions for solving computer graphics problems are given on the back of sheet 15 and on the remaining sheets in this book.

APPLICATIONS: Actual examples from industry have been used to make the problems as realistic as possible. It would be helpful if some problems were overlaid with tracing paper, traced, solved, and printed by a diazo machine as a measure of the reproducibility of your work.

COURSE ORGANIZATION: Your instructor will assign problems which are to be removed from the book, taped to your desk top, and solved as daily assignments. Your grades will be determined as indicated below:

Number

______ Daily grades ______ %

______ Tests ______ %

______ Design project ______ %

______ Written report ______ %

______ Oral report ______ %

______ Uniqueness ______ %

Optional assignments ______ %

When your graded papers have been returned, you should list your grades on page 3. This record will help you follow your progress during the course.

Record the time required to solve each problem under the heading of "minutes" in the title strip of each sheet. Keep your graded sheets in a notebook for future reference.

# Contents



REFERENCES: The numbers of the books below are used at the headings of the columns above to indicate the books and their respective chapters that can be used as references.


1. ENGINEERING DESIGN GRAPHICS (4th Edition) by James H. Earle, Reading, MA: Addison-Wesley Publishing Company, 1983.

2. DRAFTING TECHNOLOGY (Second Edition) by James H. Earle, Reading, MA: Addison-Wesley Publishing Company 1986.

3. GRAPHICS FOR ENGINEERS by James H. Earle, Reading MA: Addison-Wesley Publishing Company, 1985.

4. GEOMETRY FOR ENGINEERS by James H. Earle, Reading, MA: Addison-Wesley Publishing Company, 1984.

5. DESIGN DRAFTING by James H. Earle, Reading, MA: Addison-Wesley Publishing Company, 1972.

# Grade sheet

| Probs. | Grades | Probs. | Grades | Probs. | Grades |
|---|---|---|---|---|---|
| | | | | | |
| | | | | | |
| | | | | | |
| | | | | | |
| | | | | | |
| | | | | | |
| | | | | | |
| | | | | | |
| | | | | | |
| | | | | | |
| | | | | | |
| | | | | | |
| | | | | | |
| | | | | | |
| | | | | | |
| | | | | | |
| | | | | | |
| | | | | | |
| | | | | | |
| | | | | | |
| | | | | | |
| | | | | | |
| | | | | | |
| | | | | | |
| | | | | | |
| | | | | | |

WEEKLY TEST GRADES

TEST 1 ________

TEST 2 ________

TEST 3 ________

TEST 4 ________

TEST 5 ________

TEST 6 ________

TEST 7 ________

TEST 8 ________

TEST 9 ________

TEST 10 ________

TEST 11 ________

TEST 12 ________

TEST 13 ________

TEST 14 ________

TEST 15 ________

TEST 16 ________

TEST AVERAGE ________

MID-SEMESTER GRADE

DAILY AVERAGE ________

TEST AVERAGE ________

AVERAGE ________

FINAL GRADE

DAILY AVERAGE ________

TEST AVERAGE ________

FINAL GRADE ________

# Grades

This graph can be used to compute grades for (1) daily assignments, (2) test grades, (3) written reports, (4) oral reports, (5) uniqueness, and (6) working drawing grades. Examples of each are shown below.

1. DAILY ASSIGNMENTS & TEST GRADES
   Assignments not turned in count as zeros; however, by doing extra assignments, your grades can be improved. The example below illustrates how a grade is improved when six extra assignments have been completed.

| | |
|---|---|
| Number Assigned | 30 |
| Number Extra | 6 |
| TOTAL | 36 |

AVERAGE GRADE FOR ALL 36 — 86

$$F = \frac{\text{No. completed x 100}}{\text{No. assigned}} = \frac{36 \times 100}{30} = 120$$

From chart (F = 120) GRADE = 90

2. WRITTEN REPORTS, ORAL REPORTS, UNIQUENESS, WORKING DRAWINGS.
   Students should complete rating sheets from book for each of these areas and compute the F-values for each team member. These are used to compute each student's grade from the chart.

   If the overall team grade is 86
   the individual grades are:

| Names (N=5) | C = % | F = CN | Graph |
|---|---|---|---|
| J. D. Doe | 20% | 100 | 86 |
| H. P. Brown | 16% | 80 | 80 |
| L. O. Smith | 24% | 120 | 90 |
| R. L. Black | 20% | 100 | 86 |
| T. O. Jones | 20% | 100 | 86 |

3. DAILY GRADES

   Daily problems assigned ________
   Extra daily problems ________
   Total daily problems ________
   Avg. grade on all probs. ________
   GRADE FROM CHART ________

TEST GRADES

   Number given ________
   No. of extra assignments ________
   Total test grades ________
   GRADE FROM CHART ________

TEAM PROJECT (from chart)

   Written report ________
   Oral report ________
   Uniqueness ________
   Working drawings ________

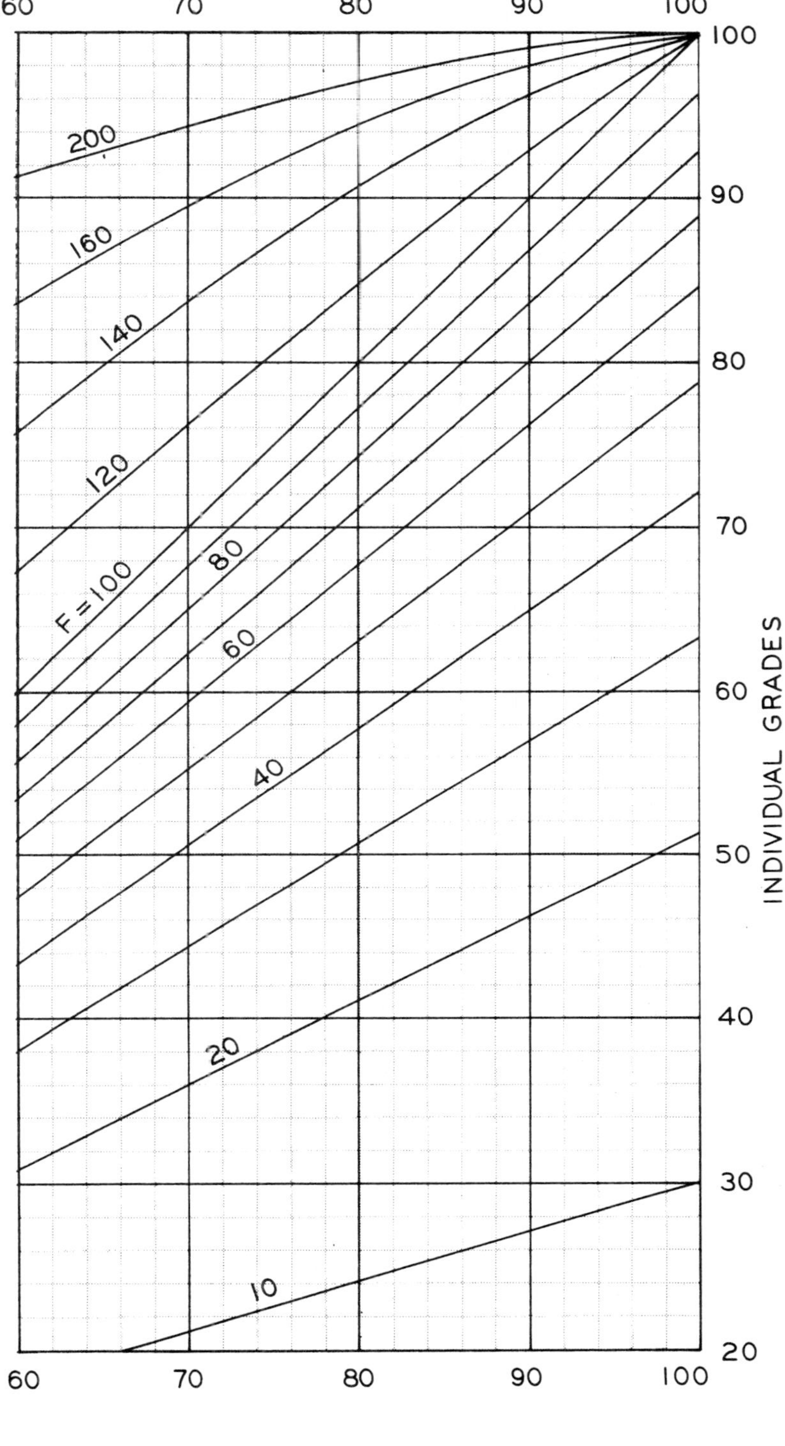


OVERALL AVERAGE [ ]

EXPECTED LETTER GRADE [ ]

# product design

These product designs are problems of the type encountered by designers of mass produced products. You are to solve the problem assigned in accordance with the specifications of your instructor. Keep all your work from the identification stage through the preparation of working drawings.

1. BABY SEAT

Design a cantilevered baby seat that could be used at home or in restaurants to suspend the child from the edge of the table in a manner similar to the illustration. A seat of this type might have a market in commercial eating establishments since they require less storage space than the conventional high chair.

Once designed, can you market this product? Do an economic analysis of of the prospects of going into production.

2. SHOPPING CADDY

Design a lightweight shopping cart that could be easily carried by the shopper. When in use, it should provide a convenient means of carrying parcels to the parking lot from the shopping center. It is important that your design be convenient and beneficial to the shopper.

Evaluate the the market potential of your design. Be sure to investigate existing designs that are on the market.

# TEAM PROJECTS

3. SIT-UP BENCH

Design a sit-up bench that can be used as an aid to physical fitness. Perhaps you can think of accessories that would make the bench serve other purposes when not being used as an exercise bench. Your design should be as economical and simple as feasible.

Survey your competition and analyze the market potential for an exercising devise of this type.

4. PRODUCT PRODUCTION

The child's toy at the right is made of wood and is held together with glue and screws. Simplify and detail this design, then concentrate on how it could be produced in numbers of 100 or 1000 per week from your garage. Determine the tools and equipment that would be needed as well as the number of employees that would be needed.

What would the toy cost to manufacture and what would the retail price be. Are there other products that your operation could produce more profitably?

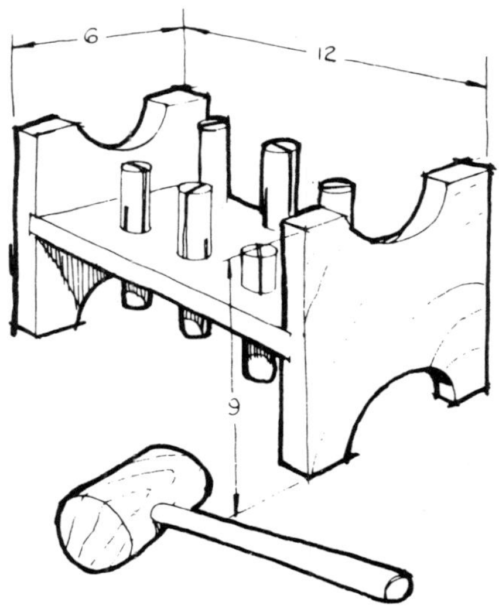

# Scheduling

## design schedule & progress record

TEAM: 2 PROJECT: PRELIMINARY DESIGN

WORK PERIODS 4 MAN HRS 20 FINISH DATE 10-2

| JOB | ASSIGNMENT | ESTIMD HOURS | ACTUAL HOURS | PER CENT COMPLETE 0 50 100 |
|---|---|---|---|---|
| 1 | TABULATE QUESTION. | 1.0 | 1.25 | |
| 2 | WRITE VENDORS | 2.0 | 1.30 | |
| 3 | PREPARE APPENDIX | 1.0 | 1.00 | |
| 4 | BEGIN PROB IDEN. | 0 | 0 | |
| 5 | PREPARE QUESTION. | 1.5 | | |
| 6 | GRAPH DATA | 3.0 | | |
| 7 | VISIT LIBRARY | 2.0 | | |
| 8 | ADMINISTER QUEST. | 2.0 | | |
| 9 | DRAFT PROB IDEN. | 2.0 | | |

The technique below is suggested as a method of scheduling activities for your team project. Good scheduling is highly essential when working as a a team.

1. List jobs that must be done as they come to mind on the DESIGN SCHEDULE & PROGRESS RECORD. These need not be listed in order, but as you think of them. Estimate the number of man-hours required to complete each job and list in the column provided. The total should not exceed the total number of hours available.

2. These jobs should then be arranged in sequence by using an ACTIVITIES NETWORK. Notice that the numbers assigned to the jobs on the DESIGN SCHEDULE & PROGRESS RECORD are the same numbers used in the ACTIVITIES NETWORK. This gives you a graphical picture of the sequence in which the jobs should be performed.

3. List the jobs on the ACTIVITIES SEQUENCE CHART, taking them from the ACTIVITIES NETWORK, to arrive at a bar graph showing when each job should be performed. Notice that the same numbers are used on the ACTIVITIES SEQUENCE CHART as were used in the previous two steps. Your ACTIVITIES SEQUENCE CHART can be used for assigning jobs to your team members. Per cent completion should be graphed on the DESIGN SCHEDULE & PROGRESS RECORD as the jobs progress toward completion.

ACTIVITIES NETWORK

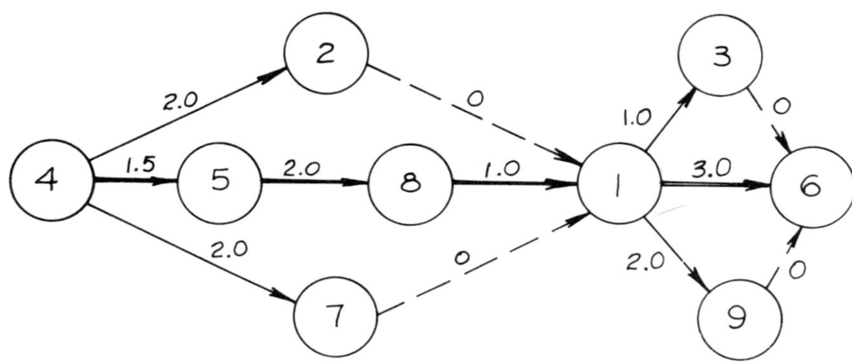


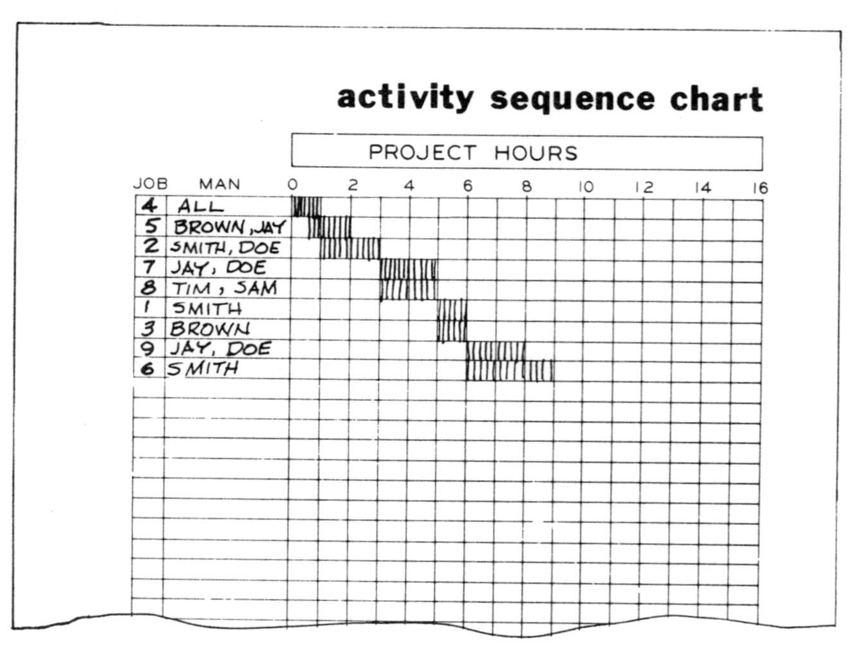

**activity sequence chart**

PROJECT HOURS

0 2 4 6 8 10 12 14 16

| JOB | MAN |
|---|---|
| 4 | ALL |
| 5 | BROWN, JAY |
| 2 | SMITH, DOE |
| 7 | JAY, DOE |
| 8 | TIM, SAM |
| 1 | SMITH |
| 3 | BROWN |
| 9 | JAY, DOE |
| 6 | SMITH |

# Design Schedule & Progress Record

TEAM ________ PROJECT ____________________________________________

WORK PERIODS ________ MAN-HOURS ________ FINISHING DATE ______ SECTION ______

| JOB | ASSIGNMENT | ESTIMD HOURS | ACTUAL HOURS | PER CENT COMPLETE<br>0 50 100 |
|---|---|---|---|---|
| | | | | |
| | | | | |
| | | | | |
| | | | | |
| | | | | |
| | | | | |
| | | | | |
| | | | | |
| | | | | |
| | | | | |
| | | | | |
| | | | | |
| | | | | |
| | | | | |
| | | | | |
| | | | | |
| | | | | |
| | | | | |
| | | | | |
| | | | | |
| | | | | |
| | | | | |
| | | | | |
| | | | | |
| | | | | |
| | | | | |
| | | | | |
| | | | | |
| | | | | |
| | | | | |
| | | | | |
| | | | | |
| | | | | |

# Activity Sequence Chart

PROJECT HOURS

| JOB | MAN | 0 | 2 | 4 | 6 | 8 | 10 | 12 | 14 | 16 |
|---|---|---|---|---|---|---|---|---|---|---|

# Progress Report

REPORT NUMBER ______

TEAM _______ PROJECT ______________________________

SECTION ______ SUBMITTED BY ____________________ DATE __________

| | 0% | 25% | 50% | 75% | 100% |
|---|---|---|---|---|---|
| **SCHEDULING** | | | | | |
| List jobs | | | | | |
| Select team leader | | | | | |
| Prepare activity sequence chart | | | | | |
| Make individual assignments | | | | | |
| Establish work guidelines | | | | | |
| **PROBLEM IDENTIFICATION** | | | | | |
| Write problem statement | | | | | |
| Determine survey data needed | | | | | |
| Write letters to vendors | | | | | |
| Estimate market needs | | | | | |
| Identify physical characteristics: sizes weights, areas, etc. | | | | | |
| Gather library data | | | | | |
| Make job assignments | | | | | |
| Write draft for written report | | | | | |
| **PRELIMINARY IDEAS** | | | | | |
| Brainstorming session | | | | | |
| Make idea sketches | | | | | |
| Adapt similar design features | | | | | |
| Write draft for written report | | | | | |
| **REFINEMENT** | | | | | |
| Prepare scale drawings | | | | | |
| Determine angles, lengths, weights, sizes, and etc. | | | | | |
| Evaluate refined designs | | | | | |
| Write draft for written report | | | | | |
| **ANALYSIS** | | | | | |
| Evaluate merits of designs | | | | | |
| Cost analysis | | | | | |
| Market analysis | | | | | |
| Weight and size analysis | | | | | |
| Strength analysis | | | | | |
| Human engineering analysis | | | | | |
| Overhead expense analysis | | | | | |
| Write draft for written report | | | | | |
| **DECISION** | | | | | |
| Select best design to implement | | | | | |
| Compare aspects of each design | | | | | |
| Make final decision | | | | | |
| **IMPLEMENTATION** | | | | | |
| Complete working drawing | | | | | |
| Graph data for report | | | | | |
| Prepare visuals for oral report | | | | | |
| Type final report | | | | | |
| Prepare final oral report | | | | | |
| Complete working model | | | | | |

# Written Report

The table below should be completed jointly by the team with only the grade column being completed by the instructor who will use the chart on page 4 and the factor F that was computed for each team member.

TEAM NO. ______ PROJECT ____________________________________________

| | NAMES | NO. (N = ) | % CONTRIBU-TION (C) | F = NC | GRADE (G) |
|---|---|---|---|---|---|
| 1. | ______________ | | ______ | ______ | ______ |
| 2. | ______________ | | ______ | ______ | ______ |
| 3. | ______________ | | ______ | ______ | ______ |
| 4. | ______________ | | ______ | ______ | ______ |
| 5. | ______________ | | ______ | ______ | ______ |
| 6. | ______________ | | ______ | ______ | ______ |
| 7. | ______________ | | ______ | ______ | ______ |
| 8. | ______________ | | ______ | ______ | ______ |

| | EVALUATION BY INSTRUCTOR 100% | Max. Value | Points Earned |
|---|---|---|---|
| 1. | Use of an appropriate cover | 2 | ______ |
| 2. | Inclusion of an evaluation sheet | 2 | ______ |
| 3. | Inculsion of a proper letter of transmittal | 2 | ______ |
| 4. | Correct title page | 2 | ______ |
| 5. | Proper table of contents | 2 | ______ |
| 6. | Sufficient introduction to the the report | 5 | ______ |
| 7. | Thoroughness in identifying the problem | 10 | ______ |
| 8. | Continuity and quality of the body the report | 10 | ______ |
| 9. | Collection and presentation of background data | 5 | ______ |
| 10. | Justification of major decisions | 5 | ______ |
| 11. | Review of costs, overhead expenses, shipping costs and similar expenses | 5 | ______ |
| 12. | Arrival at strong conclusion and recommendation | 5 | ______ |
| 13. | Sufficient number of graphs and graphics | 10 | ______ |
| 14. | Quality of graphics | 10 | ______ |
| 15. | Bibliography—form and content | 5 | ______ |
| 16. | Use of footnotes | 5 | ______ |
| 17. | Appendix—content and form | 5 | ______ |
| 18. | Form and appearance of report (spelling, punctuation, margins, typing, neatness) | 10 | ______ |
| | | 100 | |

# Working Drawings

TEAM NO. ______ PROJECT ______________________________

| NAMES | NO. (N = ) | % CONTRIBUTION (C) | F = NC | GRADE (G) |
|---|---|---|---|---|
| 1. | | | | |
| 2. | | | | |
| 3. | | | | |
| 4. | | | | |
| 5. | | | | |
| 6. | | | | |
| 7. | | | | |
| 8. | | | | |

| EVALUATION BY INSTRUCTOR | 100% | Points Earned |
|---|---|---|
| TITLE BLOCK (5 Points) | | |
| Student's name | 1 | ______ |
| Checker | 1 | ______ |
| Date | 1 | ______ |
| Scale | 1 | ______ |
| Sheet number | 1 | ______ |
| ORTHOGRAPHIC DETAILS (19 points) | | |
| Proper views | 10 | ______ |
| Spacing of views | 5 | ______ |
| Part names and numbers | 2 | ______ |
| Correct sections & conventions | 2 | ______ |
| DESIGN INFORMATION (12 Points) | | |
| Tolerances | 5 | ______ |
| Finish marks | 2 | ______ |
| Thread notes | 5 | ______ |
| DIMENSIONING (20 Points) | | |
| Proper arrowheads | 3 | ______ |
| Spacing of dimension lines | 3 | ______ |
| Adequate dimensions | 4 | ______ |
| Fillets and rounds noted | 2 | ______ |
| Hole notes | 2 | ______ |
| Inch marks omitted | 2 | ______ |
| Dimensions from best views | 4 | ______ |
| DRAFTSMANSHIP (19 Points) | | |
| Line weights | 6 | ______ |
| Lettering | 6 | ______ |
| Neatness | 3 | ______ |
| Reproduction quality | 4 | ______ |
| GENERAL NOTES (5 Points) | | |
| SI symbol | 2 | ______ |
| Third angle symbol | 2 | ______ |
| General tolerance note | 1 | ______ |
| ASSEMBLY (15 Points) | | |
| Descriptive views | 6 | ______ |
| Clarity | 3 | ______ |
| Parts list | 4 | ______ |
| Part numbers | 2 | ______ |
| PRESENTATION (5 Points) | | |
| Stapled | 2 | ______ |
| Trimmed | 1 | ______ |
| Folded | 1 | ______ |
| Grade sheet attached | 1 | ______ |
| Total | 100 | |

The table below should be completed jointly by the team with only the grade column being completed by the instructor who will use the chart on page 4 and the factor F that was computed for each team member.

# Oral Report

TEAM NO. _____ PROJECT ______________________________

| | NAMES | NO. (N = ) | % CONTRIBU-TION (C) | F = NC | GRADE (G) |
|---|---|---|---|---|---|
| 1. | ______ | | ______ | ______ | ______ |
| 2. | ______ | | ______ | ______ | ______ |
| 3. | ______ | | ______ | ______ | ______ |
| 4. | ______ | | ______ | ______ | ______ |
| 5. | ______ | | ______ | ______ | ______ |
| 6. | ______ | | ______ | ______ | ______ |
| 7. | ______ | | ______ | ______ | ______ |
| 8. | ______ | | ______ | ______ | ______ |

| EVALUATION BY INSTRUCTOR 100% | Max. value | Points earned |
|---|---|---|
| 1. Introduction of team members | 2 | ______ |
| 2. Proper dress of team members | 2 | ______ |
| 3. Statement of purpose of the presentation | 5 | ______ |
| 4. Use of visuals—point to important points, do not block screen, do not fumble, etc. | 10 | ______ |
| 5. Use of an adequate number of visual aids | 9 | ______ |
| 6. Quality of visual aids, including model | 15 | ______ |
| 7. Clear presentation of recommended design | 10 | ______ |
| 8. Presentation of alternative solutions considered | 2 | ______ |
| 9. Consideration of human factors | 5 | ______ |
| 10. Coverage of economics—manufacturing, shipping, packing, overhead, mark-up, etc. | 10 | ______ |
| 11. Presentation of an effective conclusion | 5 | ______ |
| 12. Continuity of presentation | 3 | ______ |
| 13. Poise and professionalism | 2 | ______ |
| 14. Participation of team members (perfect score if all participate) | 10 | ______ |
| 15. Use of allotted time | 10 | ______ |
| | 100 | ______ |

Additional commemts by the instructor on the back of this sheet.

# Project Uniqueness

The table below should be completed jointly by the team with only the grade column being completed by the instructor who will use the chart on page 4 and the factor F that was computed for each team member.

TEAM NO. ______ PROJECT ______________________________

| | NAMES | NO. (N = ) | % CONTRIBU-TION (C) | F = NC | GRADE (G) |
|---|---|---|---|---|---|
| 1. | | | | | |
| 2. | | | | | |
| 3. | | | | | |
| 4. | | | | | |
| 5. | | | | | |
| 6. | | | | | |
| 7. | | | | | |
| 8. | | | | | |
| | | | 100% | | |

EVALUATION BY INSTRUCTOR: Degree of uniqueness and originality in solution.

| | | Max. Value | Points Earned |
|---|---|---|---|
| 1. | Serves a needed function | 10 | ______ |
| 2. | Functions effectively | 10 | ______ |
| 3. | Needed by consumers | 10 | ______ |
| 4. | Simple, uncomplicated solution | 10 | ______ |
| 5. | Reasonable, attractive price | 10 | ______ |
| 6. | An attractive investment for marketing | 10 | ______ |
| 7. | Level of imagination and ingenuity | 10 | ______ |
| 8. | Aesthetically pleasing and attractive | 10 | ______ |
| 9. | Competition of similar products on the market | 10 | ______ |
| 10. | Degree of fulfillment of problem statement | 10 | ______ |
| | | 100 | |

Additional comments by the instructor are on the back of this sheet.

USING AN F OR HB PENCIL WITH A SLIGHTLY ROUNDED POINT, CONSTRUCT EACH LETTER IN THE SPACES PROVIDED. OBSERVE THE FORM AND PROPORTION OF EACH LETTER TO ASSIST YOU IN IMPROVING YOUR LETTERING IN FUTURE ASSIGNMENTS.

# LETTERING

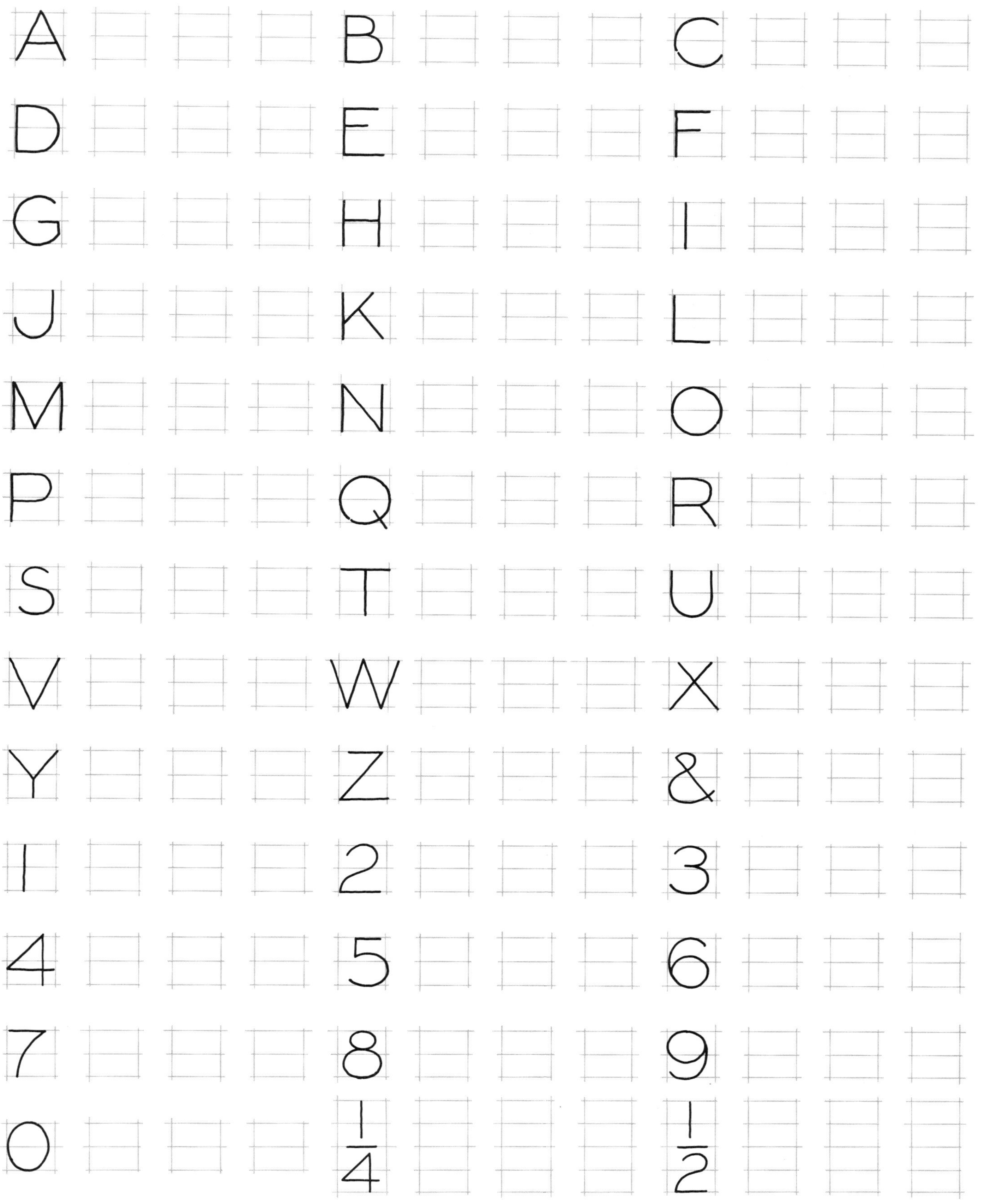




| NAME | | | MIN. | GRADE | 15 |
|---|---|---|---|---|---|
| FILE | SEC | DATE | | | |

## COMPUTER GRAPHICS PROBLEMS

GENERAL: The problems given on the backs of each sheet in this problem book are designed to be solved and constructed by computer graphics. Many microcomputer graphics software packages are available that can be used for plotting these solutions. Several types of software are: AutoCAD, VersaCAD, CADplan, Drawing Processor, CAD23, and many others.

Inclusion of problems for solution by computer provides an excellent comparison between traditional graphics and graphics by computer. Schools with limited computer equipment can rotate students from computer to drafting board to permit all principles to be covered, some by computer and some by hand with drawing instruments.

LAYOUT OF PROBLEMS: An area of 6.6" X 6.6" is given on a reduced grid at the right of each sheet. This square represents the full-size area that is available at the left of the sheet for plotting the solutions. A typical drawing area for a size A sheet for a plotter is shown in Figure 1. It is suggested that each plot have an orgin (P1) located at 0.4", 1.1" (10 mm, 30 mm) in order for the plot to fit within the available space. A grid of 0.2" matches the grid on which the problems are given.

When metric units are used, the upper right corner should be changed to 165, 165. A grid of 5 mm matches the grid given for each problem.

Where multi-pen plotters are available, you should try to use two or three pen sizes in order to maintain a good contrast between lines in keeping with good plotting practices.

PROBLEM SOLUTION: Most problems are partially laid out so they will fit in the plotting area. It would be helpful if you labeled the major points of the given problems with X- and Y-coordinates in pencil on the drawing. These coordinates would make it easier for you to plot the views on your computer's screen. The heavy-line divisions on the grid represent inches which are divided into 0.2" intervals. When using the metric system, the heavy-line divisions represent 25 mm intervals that are divided into 5 mm divisions.

In some cases you may wish to number the problems in different locations than given in the examples for better use of the space. In most examples, you will be able to use 1/8" text and numbers. However, change this size and other variables if needed.

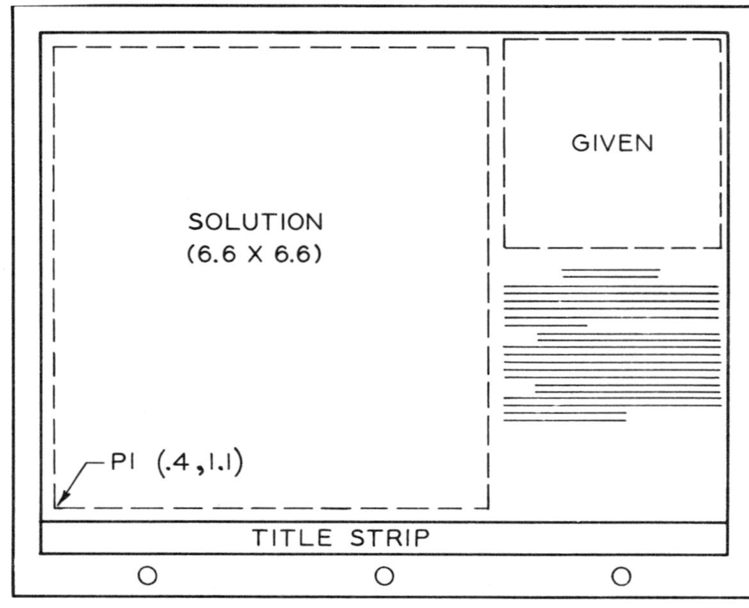


*Figure 1:*
*The layout of a typical computer graphics sheet.*

USING AN F OR HB PENCIL WITH A SLIGHTLY ROUNDED POINT, CONSTRUCT EACH LETTER IN THE SPACES PROVIDED. OBSERVE THE FORM AND PROPORTION OF EACH LETTER TO ASSIST YOU IN IMPROVING YOUR LETTERING IN FUTURE ASSIGNMENTS.

A B

C D

E F

G H

I J

K L

M N

O P

Q R

S T

U V W

X Y Z

| Graphics & Geometry | NAME | MIN. | GRADE | 16 |
|---|---|---|---|---|
| © | FILE SEC DATE | | | |

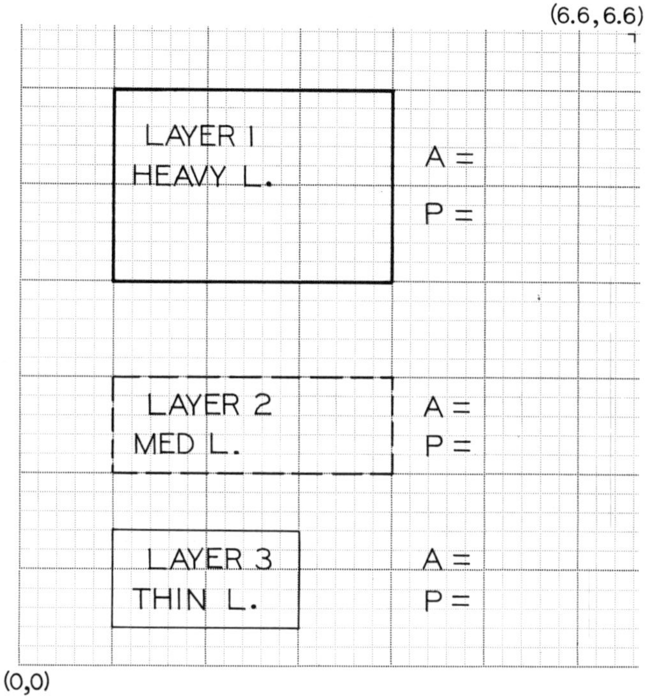

(6.6,6.6)

LAYER 1
HEAVY L.

A =
P =

LAYER 2
MED L.

A =
P =

LAYER 3
THIN L.

A =
P =

(0,0)

## LAYERS AND PENS

Plot the rectangles using the layers and pens that are specified. The rectangle on layer 2 is to be drawn with hidden (dashed) lines.

Find the areas (A) and perimeters (P) of each and note them to the right of each rectangle.

# USE OF INSTRUMENTS

MAKE A DOUBLE-SIZE INSTRUMENT DRAWING OF THE SKETCH OF THE TRANSMITTER CARRIER FREQUENCY CIRCUIT GIVEN BELOW. START IN THE UPPER LEFT CORNER AT POINT A AND USE THE FULL-SIZE CIRCUIT COMPONENTS THAT ARE GIVEN. LABEL THE COMPONENTS AS GIVEN IN THE SKETCH OF THE CIRCUIT.

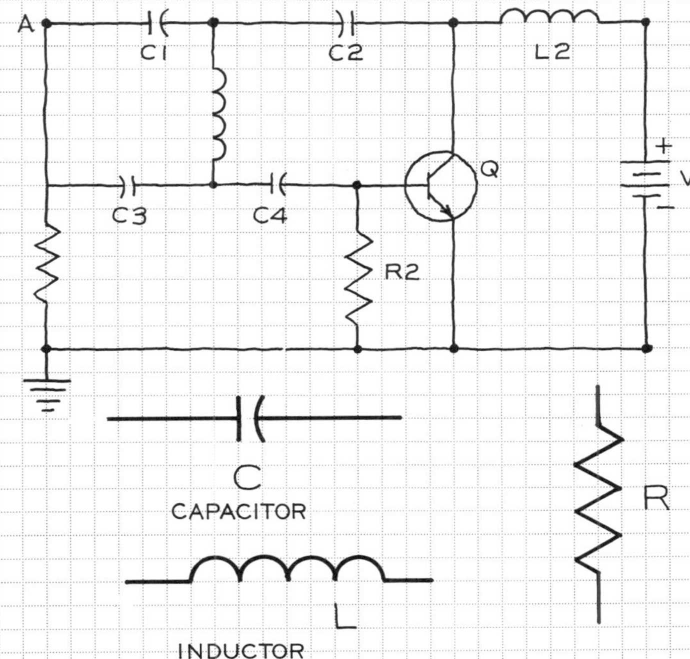


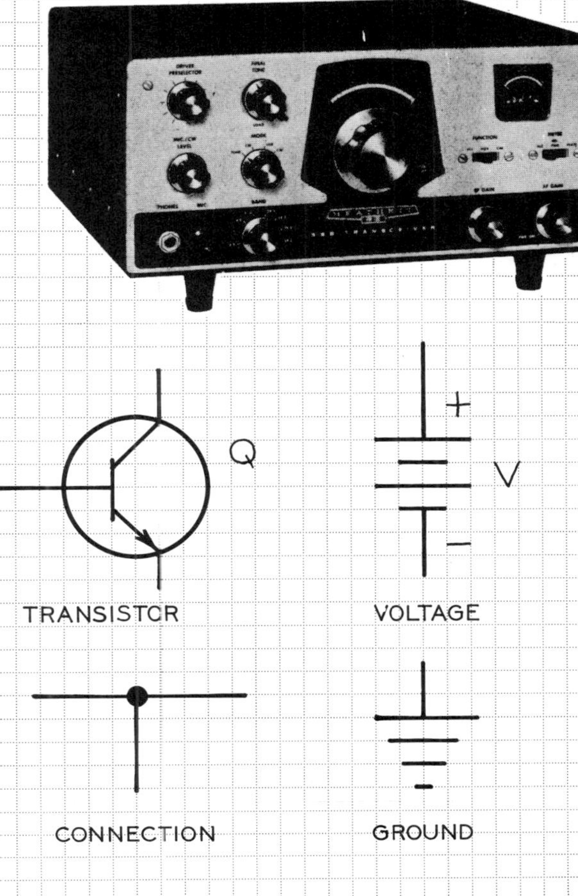

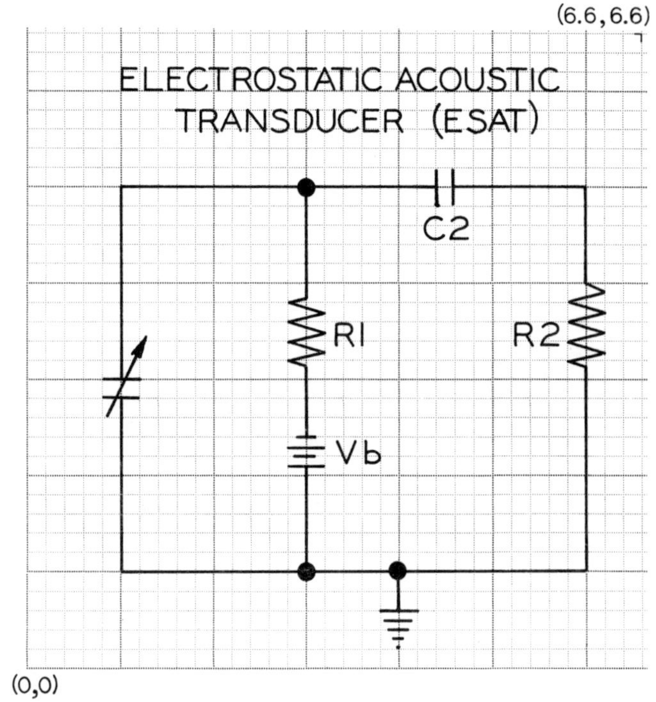

(6.6,6.6)

ELECTROSTATIC ACOUSTIC TRANSDUCER (ESAT)

C2

R1

R2

Vb

(0,0)

## CIRCUIT PLOTTING

Plot the circuit for the Electrostatic Acoustic Transducer (ESAT). The general proportions of the symbols can be determined from the grid. The dimensions of the circuit do not have to exact, but approximately similar to the example shown.

Use commands such as COPY and MOVE to plot the resistor symbols. (These commands may be different if you are using a program other than AUTOCAD.)

# SCALES

## ARCHITECTS' SCALE

SOLVE PROBLEMS 1 THROUGH 6 BY MEASURING LENGTHS A, B AND C ON FIGURE 1 USING THE SCALES GIVEN BELOW. LETTER EACH DIMENSION IN THE APPROPRIATE COLUMN. MEASURE TO THE NEAREST SMALL DIVISION ON THE SCALE USED.

| | SCALE | A | B | C |
|---|---|---|---|---|
| 1 | 1 = 1'-0 | | | |
| 2 | $\frac{1}{2}$ = 1'-0 | | | |
| 3 | $\frac{3}{4}$ = 1'-0 | | | |
| 4 | $\frac{3}{8}$ = 1'-0 | | | |
| 5 | $1\frac{1}{2}$ = 1'-0 | | | |
| 6 | $\frac{1}{4}$ = 1'-0 | | | |

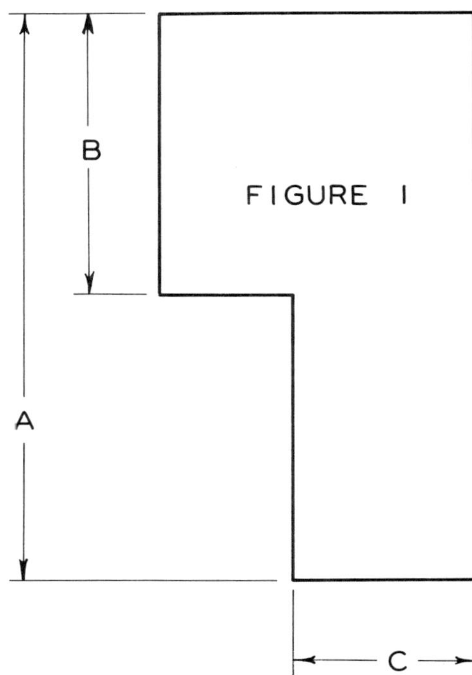


## ENGINEERS' SCALE

SOLVE PROBLEMS 7 THROUGH 12 BY MEASURING LENGTHS D, E AND F ON FIGURE 2 USING THE SCALES GIVEN BELOW. LETTER EACH DIMENSION IN THE APPROPRIATE COLUMN. ESTIMATE YOUR MEASUREMENT TO THE NEAREST HALF OF THE SMALLEST SCALE DIVISION.

| | SCALE | D | E | F |
|---|---|---|---|---|
| 7 | 1 = 10' | | | |
| 8 | 1 = 50' | | | |
| 9 | 1 = 400' | | | |
| 10 | 1 = 30' | | | |
| 11 | 1 = 200' | | | |
| 12 | 1 = 600' | | | |

## ANGLES

SOLVE PROBLEMS BELOW BY MEASURING TO THE NEAREST PRACTICAL DIMENSION THE INTERIOR AND DEFLECTION ANGLES TO THE RIGHT AT 1, 2, 3 AND 4 ON FIGURE 2 (DEFLECTION ANGLES MEASURED IN CLOCKWISE DIRECTION). LETTER EACH DIMENSION IN THE APPROPRIATE COLUMN. TOTAL EACH COLUMN.

| | | INTERIOR ∠'S | DEFLECTION ∠'S |
|---|---|---|---|
| 13 | 1 | | |
| 14 | 2 | | |
| 15 | 3 | | |
| 16 | 4 | | |
| 17 | TOTAL | | |

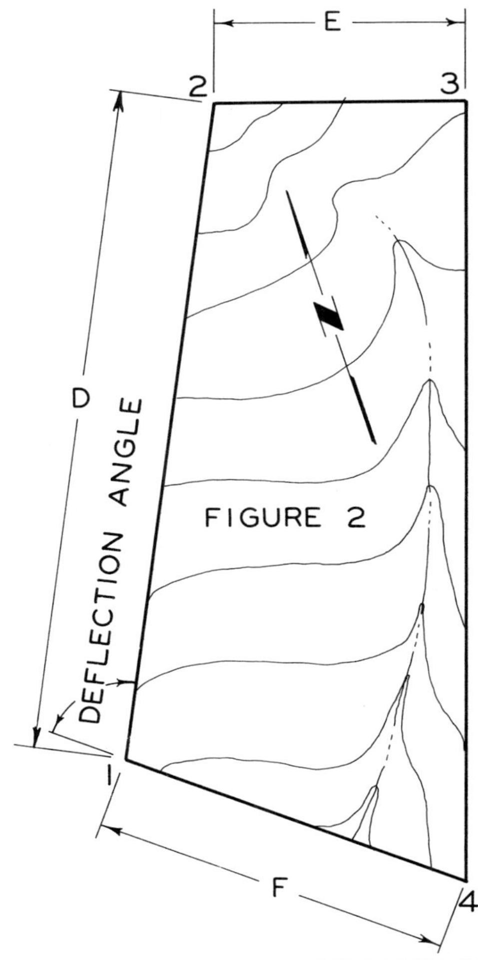

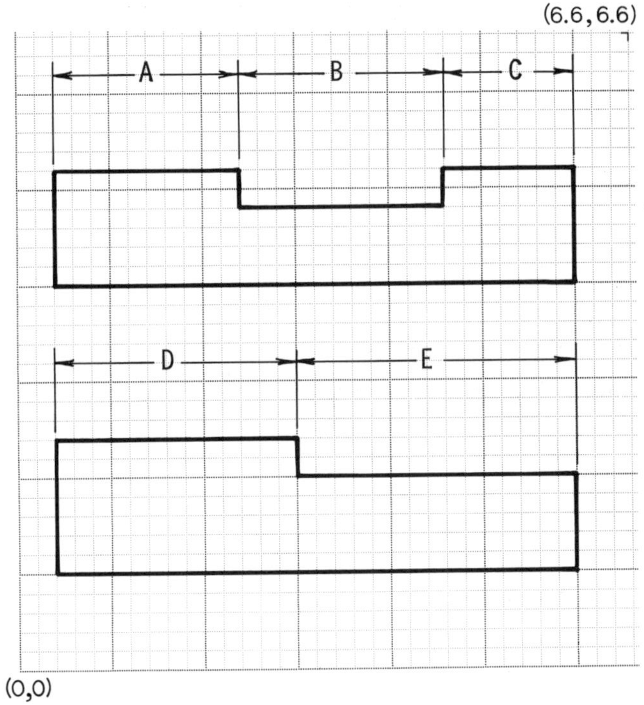

(6.6,6.6)

A B C

D E

(0,0)

## ARCHITECTS' & ENGINEERS' SCALE

Solve the problem assigned.

1. Plot and dimension the parts using the architects' scale where each square is equal to 6 inches.

2. Plot and dimension the parts using the engineers' scale where each square is equal to 0.5 feet.

# METRIC SCALES

MEASURE THE DIMENSIONS ON THE PART (FIG. 1), USING METRIC SCALES INDICATED. LETTER THE VALUES UNDER THE HEADINGS: A, B, & C. NOTICE THAT THE MEASUREMENTS UNDER A WILL BE IN MILLIMETERS, UNDER B IN CENTIMETERS AND UNDER C IN METERS.

CENTIMETERS
0 1 2 3 4 5 6 7 8 9
0 1 2 3
DECIMAL INCHES

| LINEAR | MILLIMETERS A (mm) | CENTIMETERS B (cm) | METERS C (m) |
|---|---|---|---|
| **1** 1 : 1 | | | |
| **2** 1 : 3 | | | |
| **3** 1 : 40 | | | |
| **4** 1 : 50 | | | |
| **5** 1 : 200 | | | |
| **6** 1 : 60 | | | |

A
B
C
FIG 1

USING A CENTIMETER SCALE, DETERMINE DIMENSIONS H, W, & D FROM THE PICTORIAL AND LETTER THE ANSWERS UNDER THE APPROPRIATE HEADINGS.
1 LITER = 1000 CUBIC CENTIMETERS

| VOLUME | H (mm) | X | W (mm) | X | D (mm) | = | LITERS |
|---|---|---|---|---|---|---|---|
| **7** 1 : 1 | | X | | X | | = | |
| **8** 1 : 30 | | X | | X | | = | |
| **9** 1 : 40 | | X | | X | | = | |

D
H
W
VOL = ?

USING THE DENSITIES BELOW AND THE VOLUMES FOUND IN PROBLEMS 7-9, DETERMINE THE WEIGHTS OF THE VOLUMES IN GRAMS.

| MASS | | | LITERS | | | KILO GRAMS |
|---|---|---|---|---|---|---|
| **10** WATER | (1000 GRAMS / L) | X | | (PROB 7) | = | |
| **11** STEEL | (7850 GRAMS / L) | X | | (PROB 8) | = | |
| **12** CONCRETE | (2200 GRAMS / L) | X | | (PROB 9) | = | |

A FAHRENHEIT SCALE IS GIVEN FROM FREEZING (32°) TO BOILING (212°). BY GEOMETRIC CONSTRUCTION CALIBRATE THE SCALE SHOWING INTERVALS OF 10° FROM 0° TO 100° ON THE LOWER SIDE TO REPRESENT CENTIGRADE.

32 212
0 20 40 60 70 80 80 90 100 100

TEMPERATURE

**13**

° FAHRENHEIT
32 212
30 40 60 80 100 120 140 160 180 200 220
0 100
° CENTIGRADE

| **Graphics & Geometry** © | NAME / FILE SEC DATE | MIN. | GRADE | 19 |
|---|---|---|---|---|

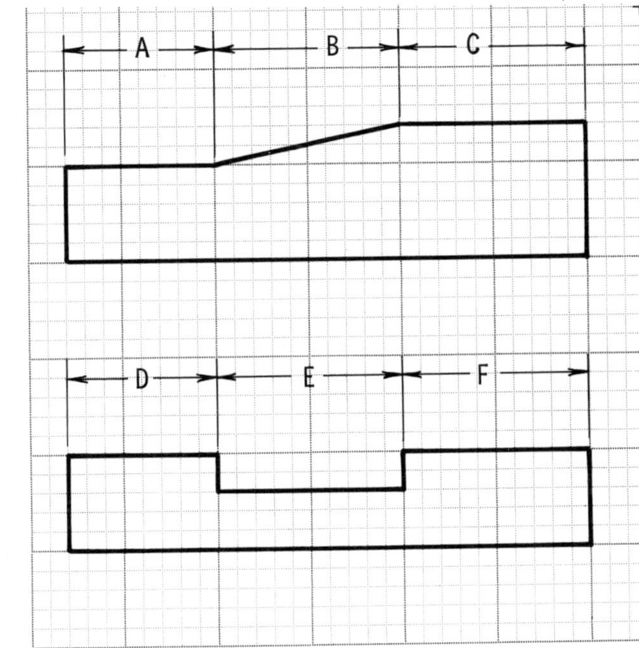


## METRIC SCALES

Plot and dimension the full-size parts in millimeters where each square is equal to 5 mm.

Experiment with various scales. For example, set limits at 0,0 and 333, 333, and let each square equal 10 mm. This has the same effect as drawing the parts half size.

Also, you may select different plotting scales to vary the size of finished drawings.

# LINE STANDARDS

COURTESY OF THOMSON INDUSTRIES, INC.

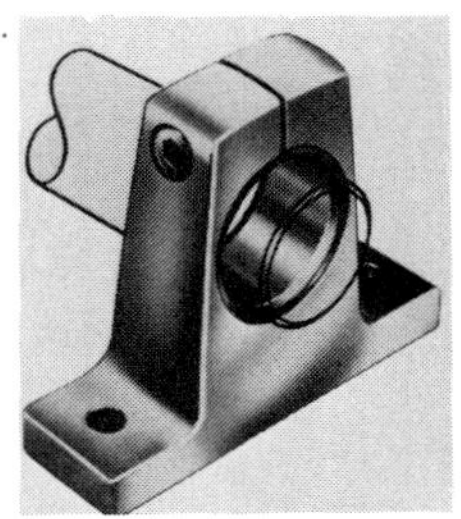

USING THE SUGGESTED PENCIL GRADES, MAKE A COPY OF THE SHAFT SUPPORT. OMIT INSTRUCTIONAL NOTES.

1. MAKE A FULL-SIZE COPY BEGINNING AT THE GIVEN CORNER.
2. OVERLAY THIS DRAWING WITH AN 8-1/2" X 11" SHEET OF VELLUM FROM THE BACK OF THE BOOK AND MAKE A TRACING THAT CAN BE REPRODUCED BY THE DIAZO PROCESS.

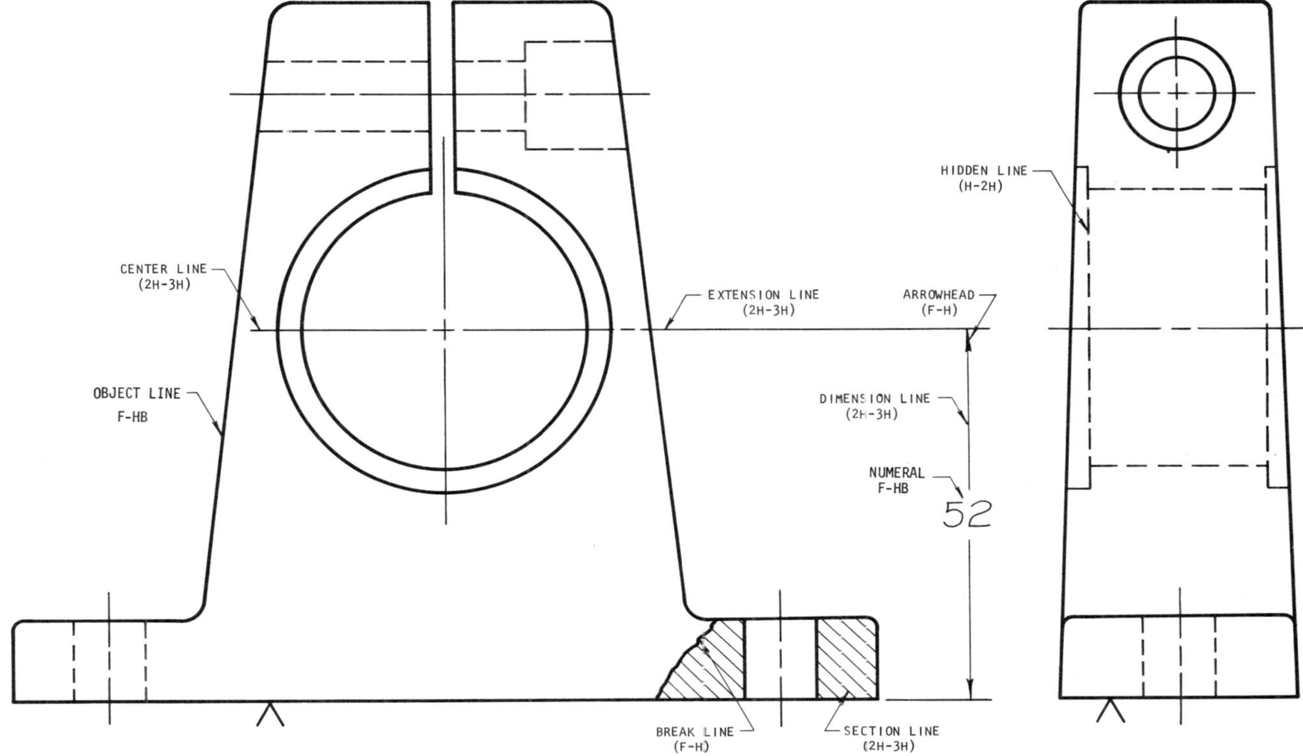


START HERE

| **Graphics & Geometry** © | NAME<br>FILE SEC DATE | MIN. | GRADE | 20 |
|---|---|---|---|---|

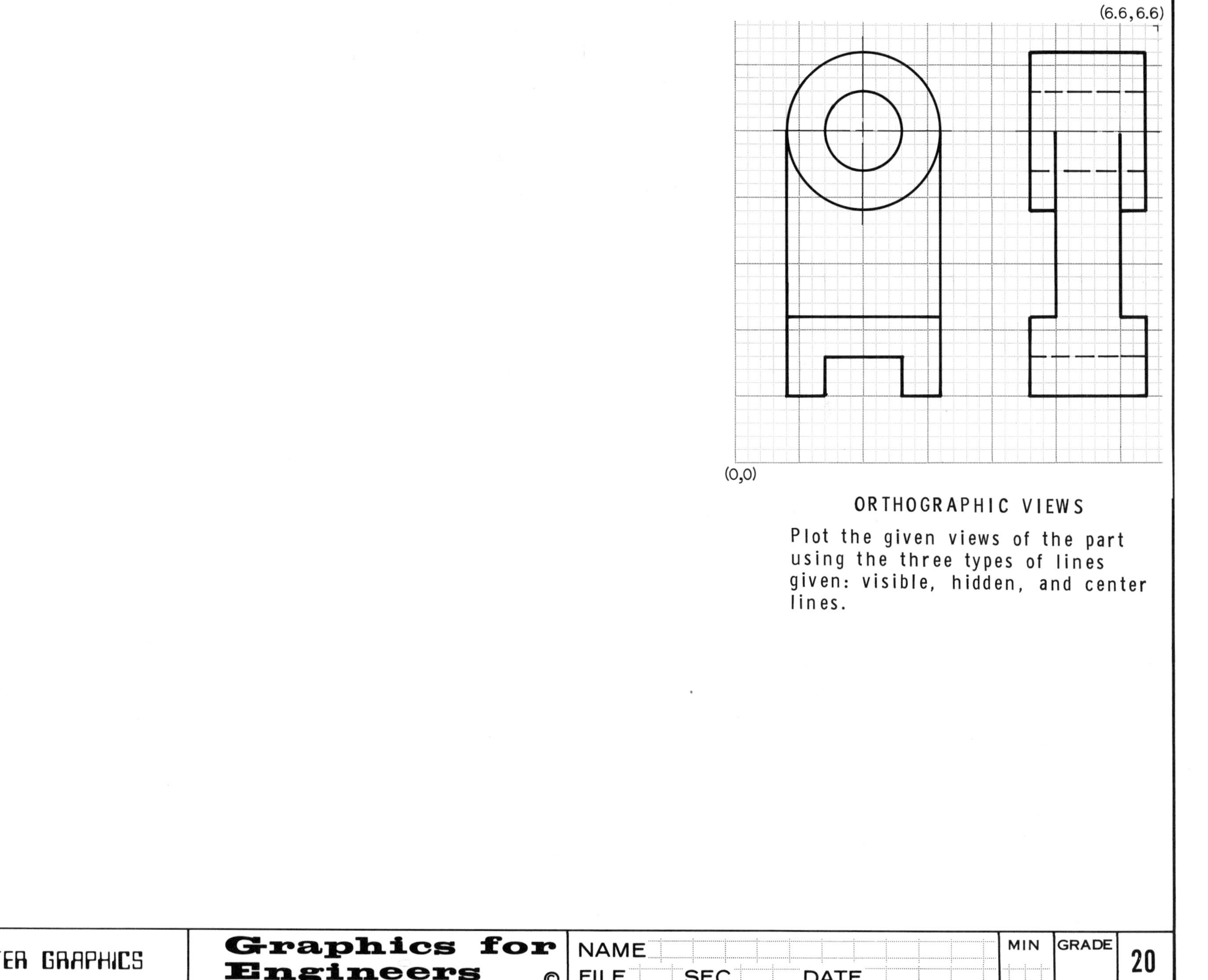


## ORTHOGRAPHIC VIEWS

Plot the given views of the part using the three types of lines given: visible, hidden, and center lines.

| COMPUTER GRAPHICS | Graphics for Engineers © | NAME<br>FILE SEC DATE | MIN | GRADE | 20 |
|---|---|---|---|---|---|

CONSTRUCT ARCS TANGENT TO THE ROAD SEGMENTS USING THE CENTER-LINE ARCS SPECIFIED IN THE TABLE. EXTEND THE ROAD SEGMENTS AS NECESSARY. SHOW CONSTRUCTION, MARK POINTS OF TANGENCY, AND COMPLETE THE MISSING SEGMENTS OF THE ROADS USING LINES THAT MATCH THE GIVEN LINES. (NOTE: THERE WILL BE A TANGENT CURVE ON BOTH SIDES OF THE GIVEN ARC IN PROBLEM 3.

# TANGENCIES

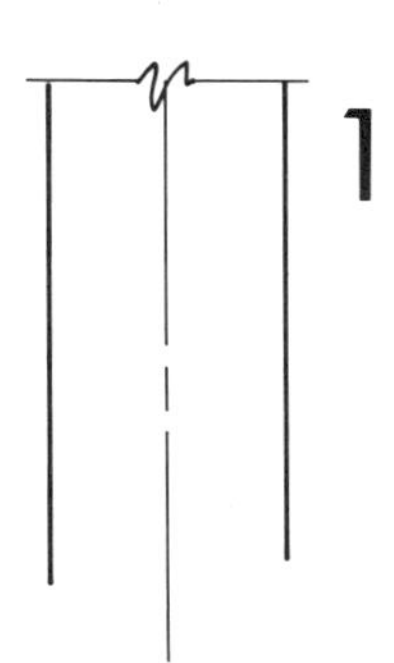

1

| PROB | FT | m |
|---|---|---|
| 1 | 70 | 21 |
| 2 | 50 | 15 |
| 3 | 55 | 17 |
| 4 | 80 | 24 |

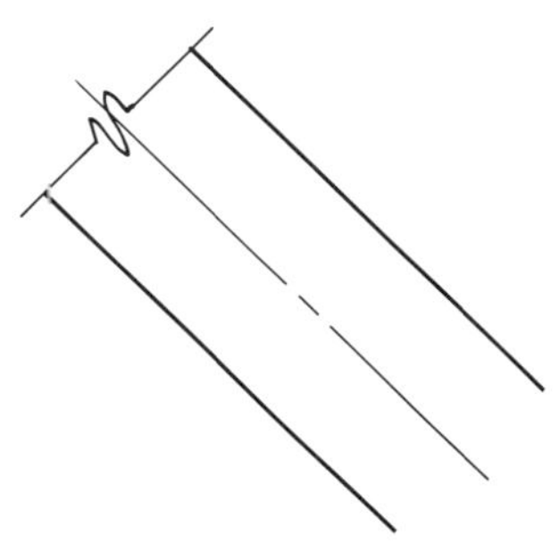

2

3

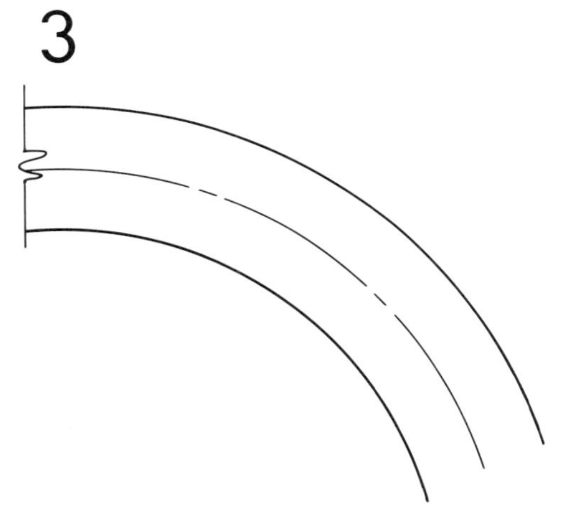

C

SCALE: 1 = 50
METRIC 1:600

4

C

| © Graphics & Geometry | NAME<br>FILE SEC DATE | MIN. | GRADE | 21 |
|---|---|---|---|---|

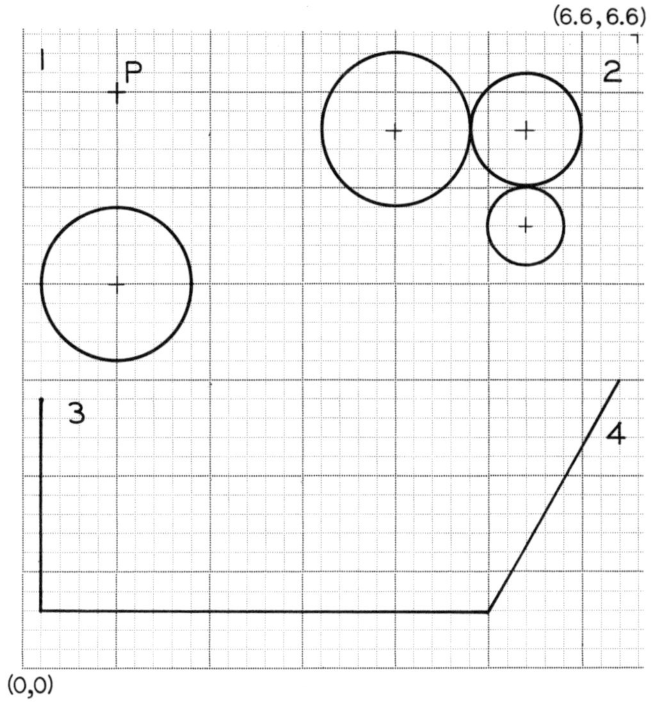

(6.6,6.6)

1 P

2

3

4

(0,0)

## GEOMETRIC CONSTRUCTION

1. Plot lines that are tangent to the circle from point P and locate points of tangency.
2. Locate the points of tangency between the circles.
3. Construct an arc of 1.40" R tangent to the perpendicular lines and mark points of tangency.
4. Same as problem 3 except the lines are not perpendicular.

# TANGENCIES

1 CONSTRUCT AN ARC OF 1.5" (38 mm) R TANGENT AND CONCAVE TO THE GIVEN ARCS. EXTEND THE GIVEN ARCS AS NECESSARY. SHOW CONSTRUCTION AND MARK POINTS OF TANGENCY.

2 CONSTRUCT AN ARC OF 2.5" (64 mm) R TANGENT TO AND CONVEX WITH THE GIVEN ARCS. EXTEND THE GIVEN ARCS AS NECESSARY. SHOW CONSTRUCTION AND MARK POINTS OF TANGENCY.

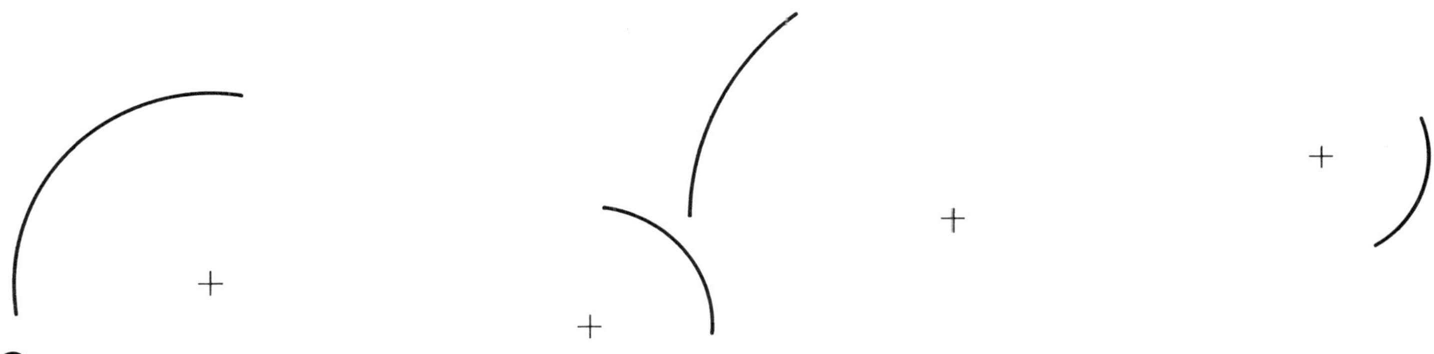

3 USING THE GIVEN SPECIFICATIONS, CONSTRUCT THE HIGHWAY ABOUT ITS CENTER LINES. THE RADII ARE GIVEN FROM THE CENTER LINES. SHOW CONSTRUCTION AND MARK POINTS OF TANGENCY.

40'

A - 40.0'(12m)R

C - GIVEN R

B - 60.0 (18m)R

D - GIVEN R

40'

SCALE: 1=40'
SCALE: 1:500

COURTESY TEXAS TRANSPORTATION INSTITUTE

E - 80.0'(24m)R

| Graphics for Engineers © | NAME | | | MIN. | GRADE | 22 |
|---|---|---|---|---|---|---|
| | FILE | SEC | DATE | | | |

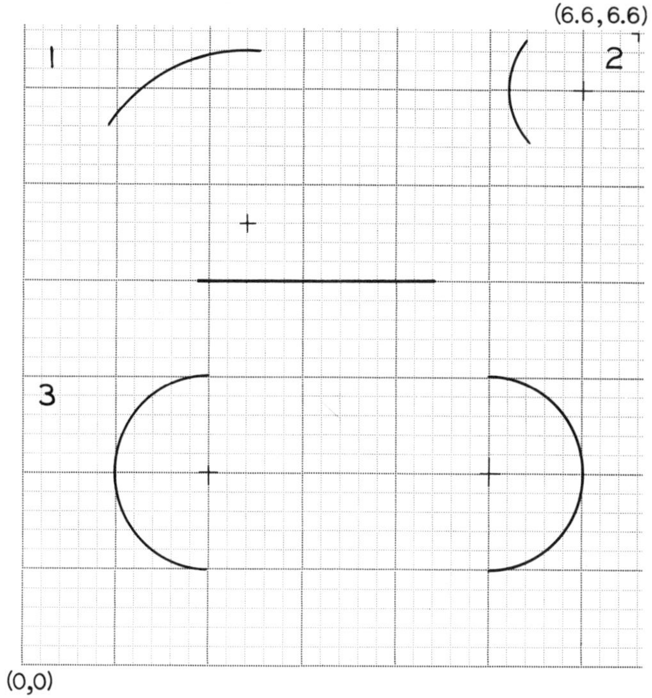

(6.6,6.6)

1

2

3

(0,0)

GEOMETRIC CONSTRUCTION

1 & 2. Using radii of 0.8", construct arcs tangent to the arcs and line.

3. Using a radius of 1.20", construct arcs tangent to the two arcs on the upper and lower sides of the part.

*The arcs and lines in all problems may need to be extended.

| COMPUTER GRAPHICS | **Graphics for Engineers** © | NAME<br>FILE SEC DATE | MIN | GRADE | 22 |
|---|---|---|---|---|---|

# SIX-VIEW SKETCHING

GIVEN THE FRONT AND RIGHT SIDE VIEWS OF THE RIGHT ANGLE BLOCKS SHOWN BELOW, SKETCH THE REMAINING VIEWS, LABEL THEM AND PLACE THE OVERALL DIMENSIONS OF W, H AND D ON THE APPROPRIATE VIEWS. WHEN COMPLETED, CUT ALONG THE DASHED OUTSIDE LINES AND FOLD ALONG THIN LINES TO FORM A SIX VIEW MODEL.

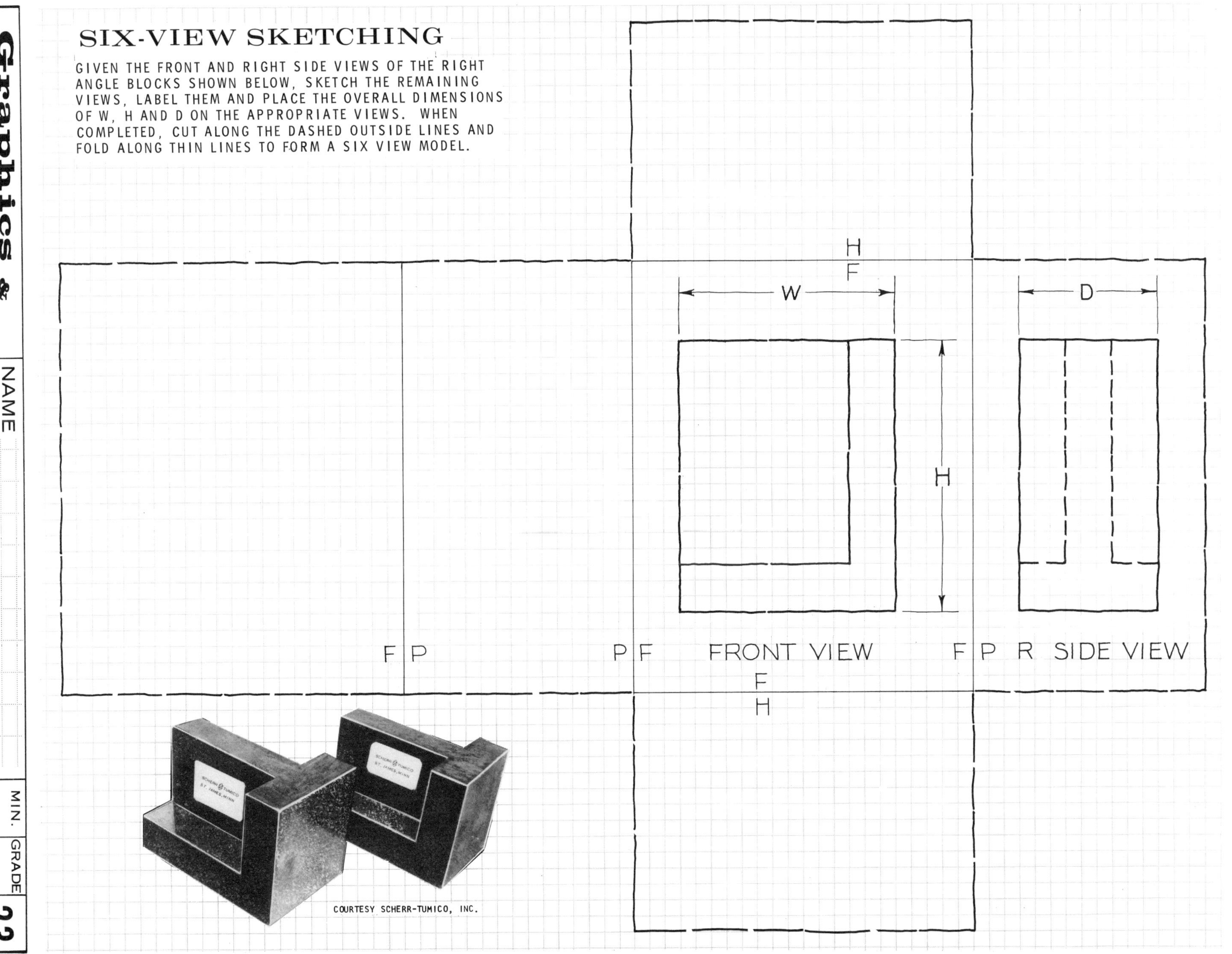


COURTESY SCHERR-TUMICO, INC.

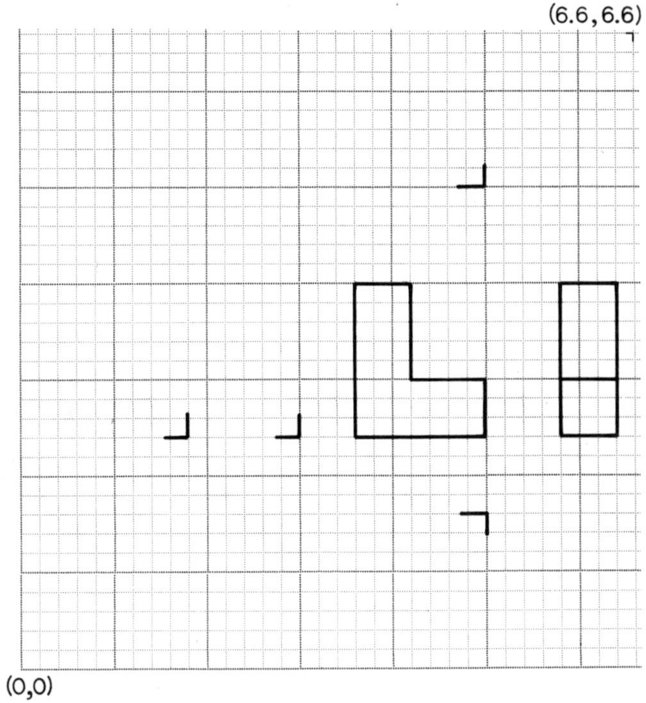

(6.6,6.6)

(0,0)

## SIX-VIEW DRAWINGS

Complete the six principal views of the object given in the views above. Label the views with 1/8" letters. Use dimensions of W, D, and H that are 0.10" high to give the overall dimensions.



NAME
FILE SEC DATE
MIN GRADE

# THREE-VIEW SKETCHING

1

PROBLEM 1: DOUBLE THE DIMENSIONS IN THE ISOMETRIC PICTORIAL, AND SKETCH THREE ORTHOGRAPHIC VIEWS OF THE SUPPORT AT THE LEFT. LABEL THE VIEWS AND GIVE OVERALL DIMENSIONS AS H, W, & D.

PROBLEM 2: USING THE FULL-SIZE DIMENSIONS GIVEN, SKETCH THREE ORTHOGRAPHIC VIEWS OF THE TOOL HOLDER. LABEL THE VIEWS AND GIVE OVERALL DIMENSIONS USING THE TRUE NUMERALS.

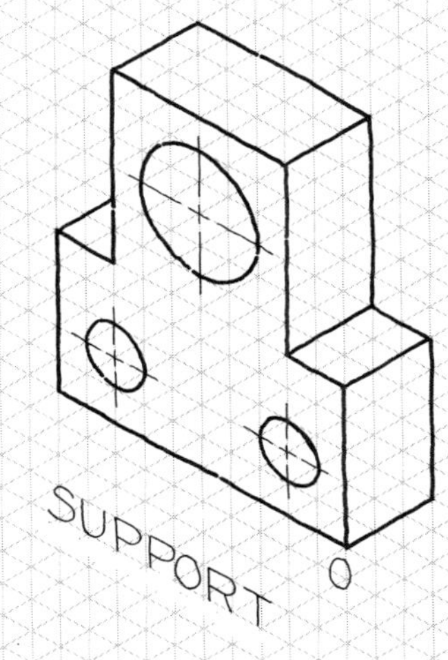


2

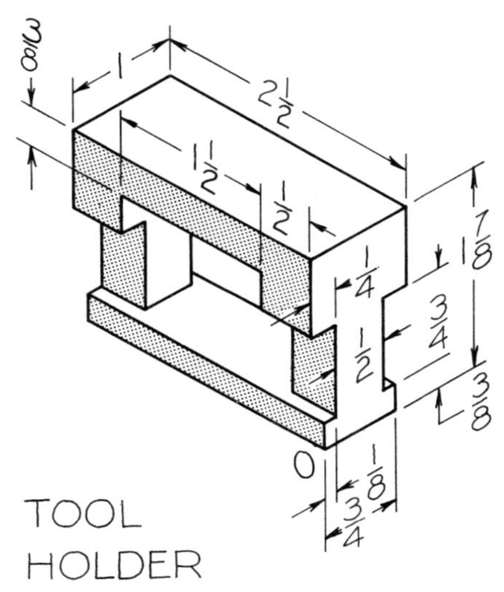




©

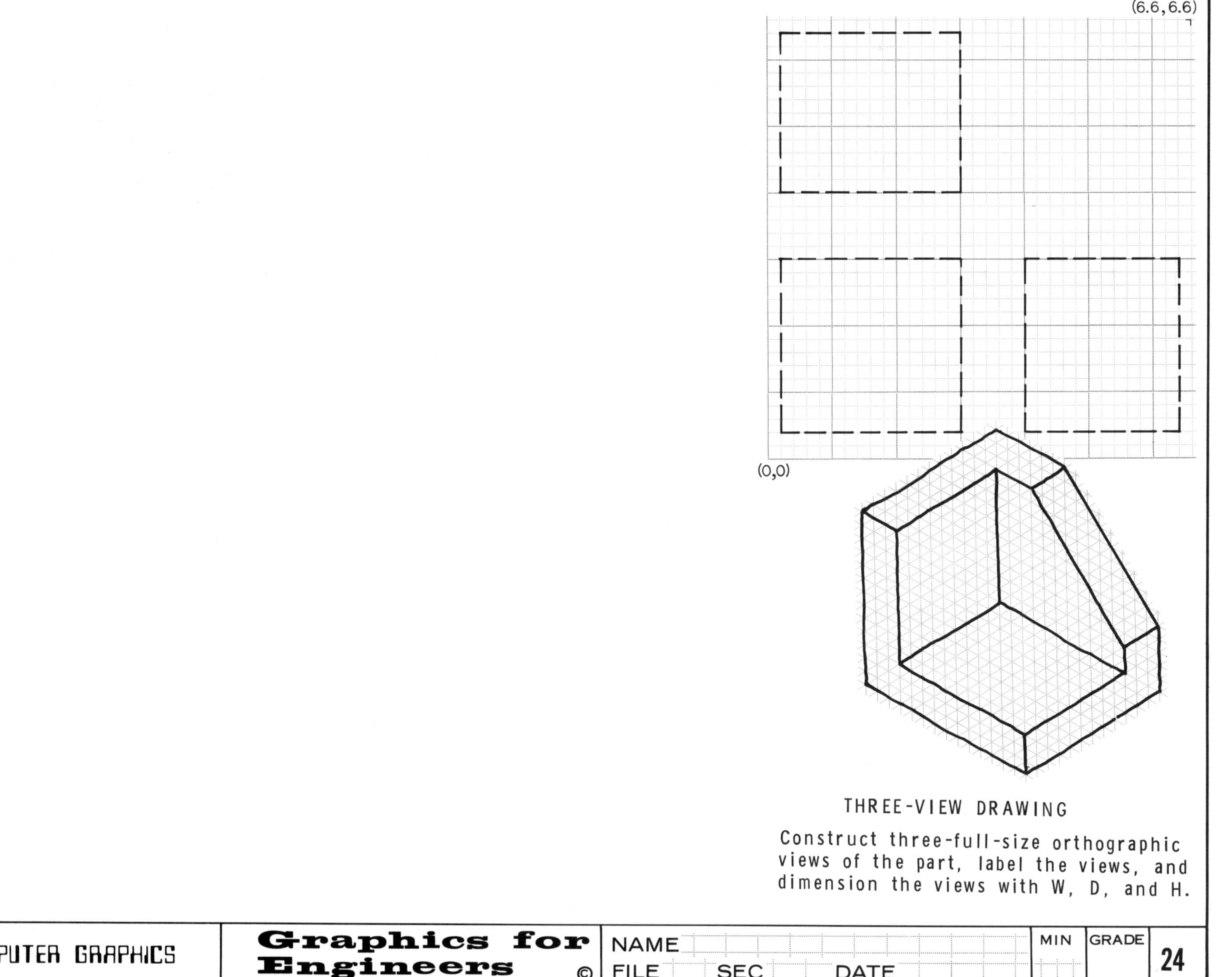


## THREE-VIEW DRAWING

Construct three-full-size orthographic views of the part, label the views, and dimension the views with W, D, and H.

| COMPUTER GRAPHICS | **Graphics for Engineers** © | NAME<br>FILE SEC DATE | MIN | GRADE | 24 |
|---|---|---|---|---|---|

SUPPLY THE MISSING LINES TO COMPLETE ALL VIEWS.

1

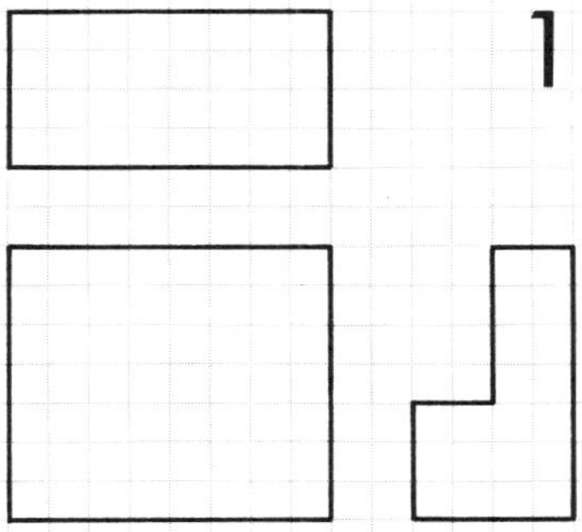

2

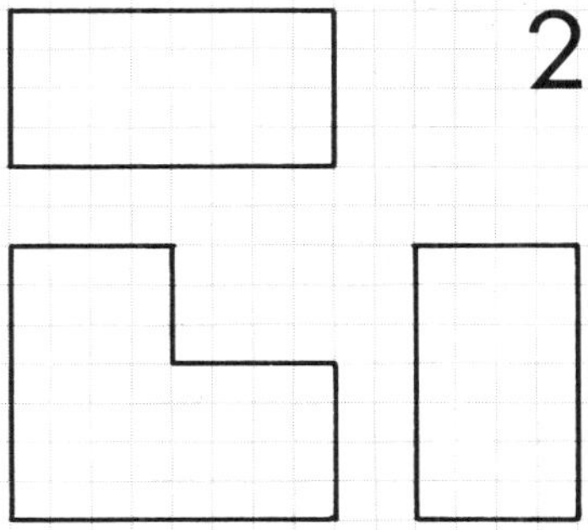

3

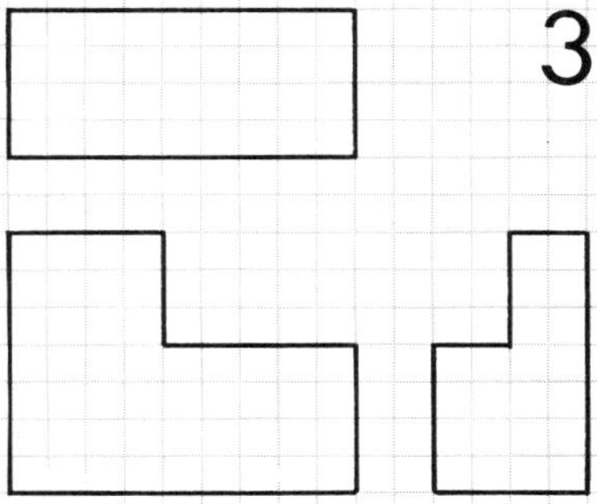

4

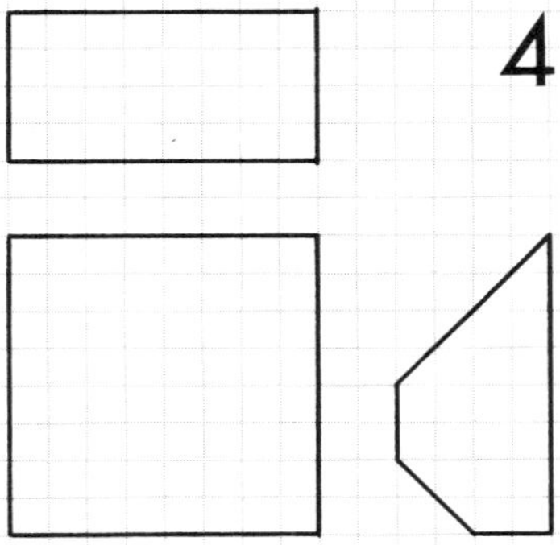

5

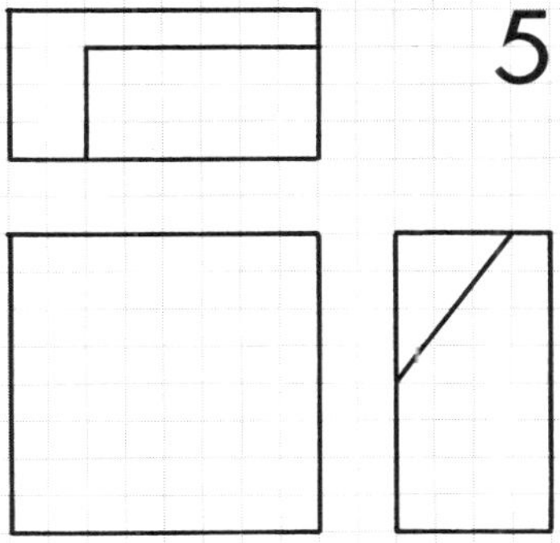

6

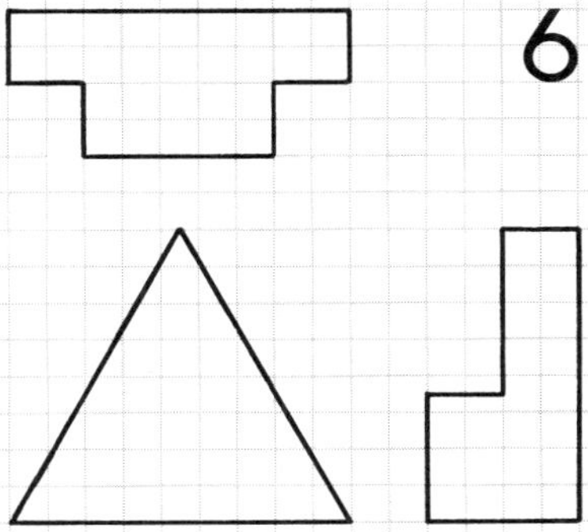

7

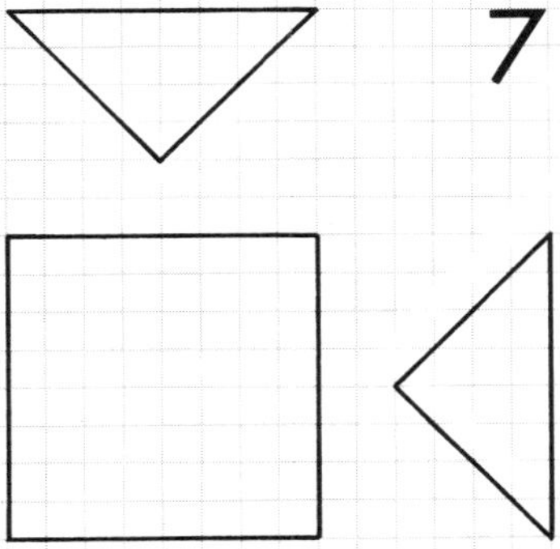

8

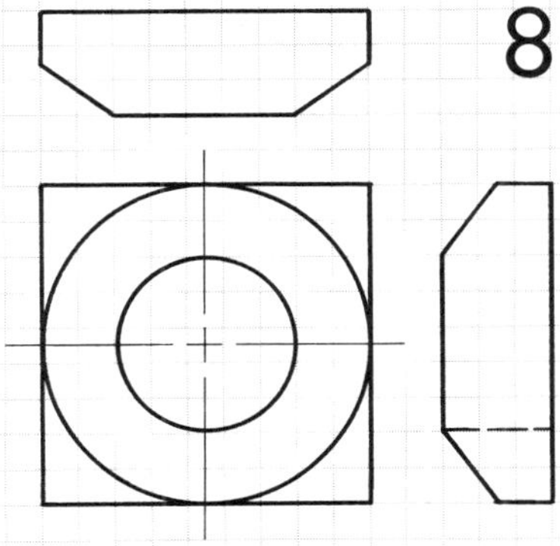

9

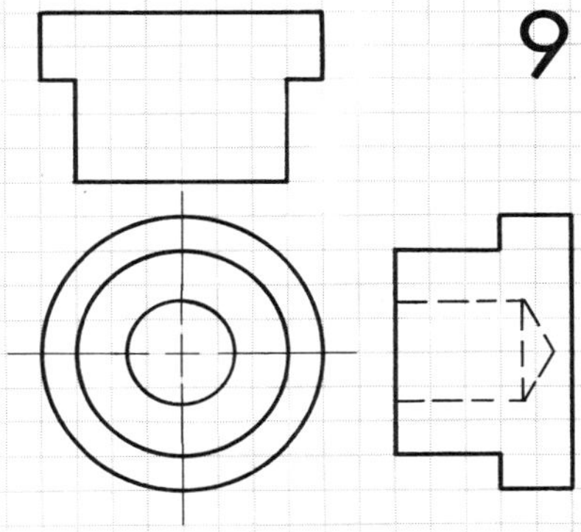

10

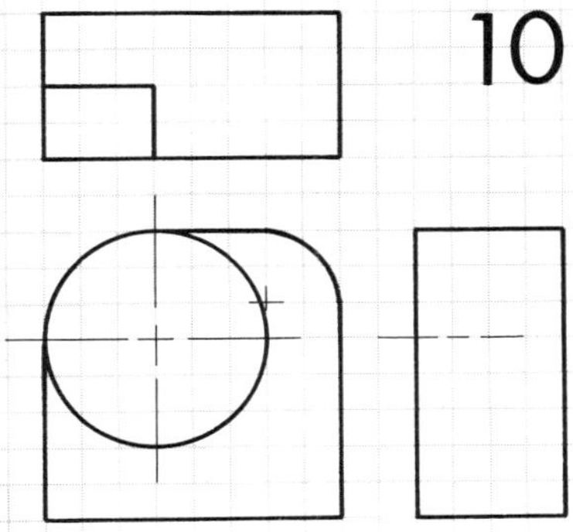

11

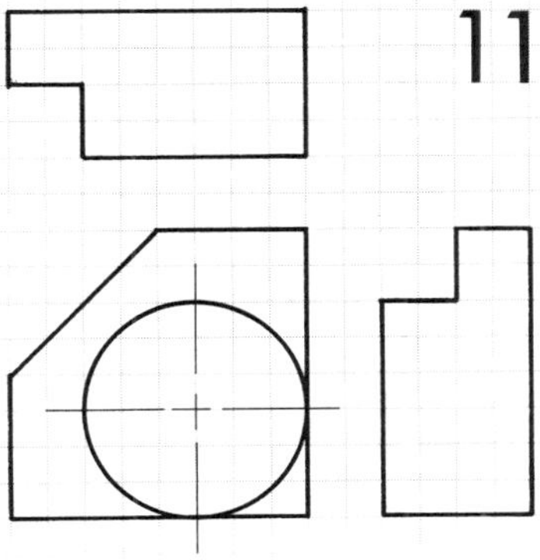

12

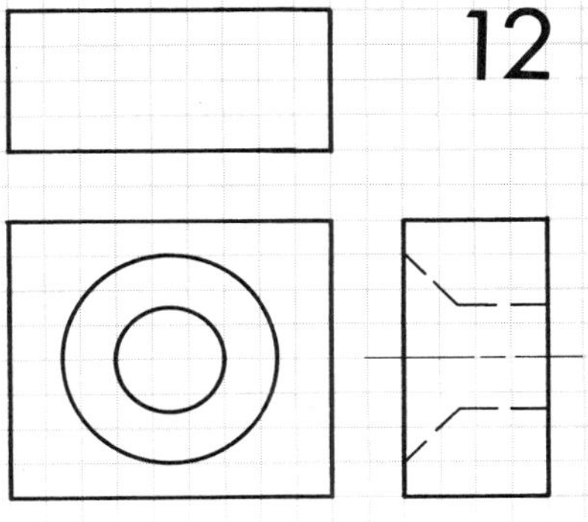

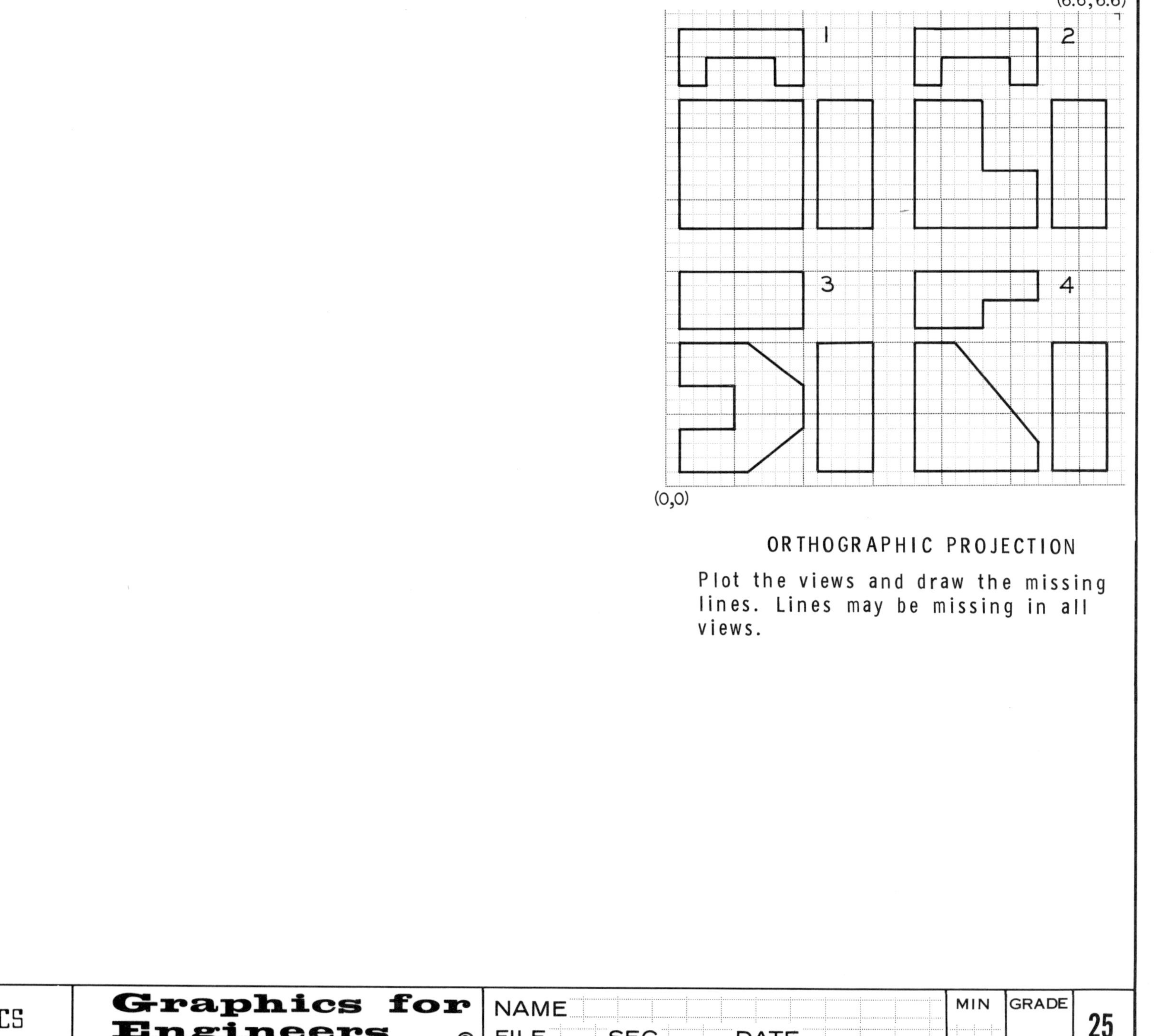


## ORTHOGRAPHIC PROJECTION

Plot the views and draw the missing lines. Lines may be missing in all views.

# DESIGN MODIFICATION

REDSIGN THE BRACKET AT THE RIGHT ACCORDING TO THE FOLLOWING SPECIFICATIONS: THE CYLINDER HAS A DIAMETER OF 2.00" (51 mm) AND IT EXTENDS 2.25" (57 mm) ABOVE THE TOP OF THE BASE. A 1.25" (32 mm) DIA HOLE IS BORED THROUGH THE CYLINDER. MODIFY THE BASE TO BE CIRCULAR INSTEAD OF SQUARE WITH A DIAMETER OF 4.00" (100 mm). FOUR EQUALLY SPACED HOLES ARE TO BE DRILLED THROUGH THE BASE. THE CENTERS OF THE 0.38" (9mm) HOLES ARE TO BE LOCATED ON A 3.00" (77 mm) DIAMETER CIRCLE OF CENTERS.

SKETCH (FREEHAND) THE FRONT AND TOP VIEWS OF THE NEW DESIGN.

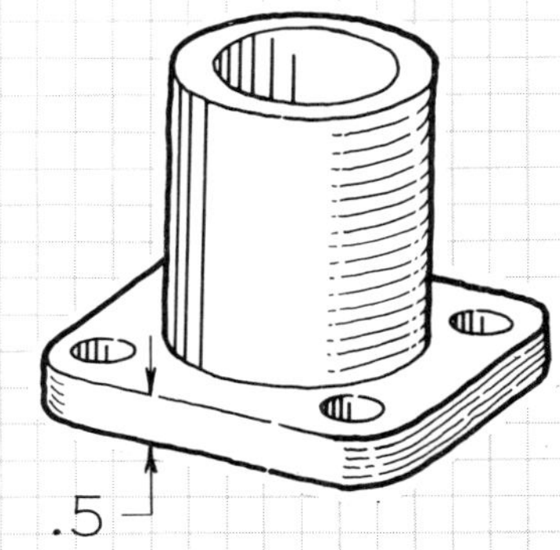


FULL SIZE



©

## DESIGN MODIFICATION

Plot the views of the Bracket in accordance with the redesign specifications given on the opposite side of this sheet.

# ISOMETRIC SKETCHING

**1** SKETCH THE ISOMETRIC PROJECTION OF THE FRONT, TOP AND RIGHT-SIDE VIEWS OF THE TOOL HOLDER SUPPORT PROJECTED TO THE RESPECTIVE PRINCIPAL PLANES OF PROJECTION. POINT O IS ALREADY PROJECTED.

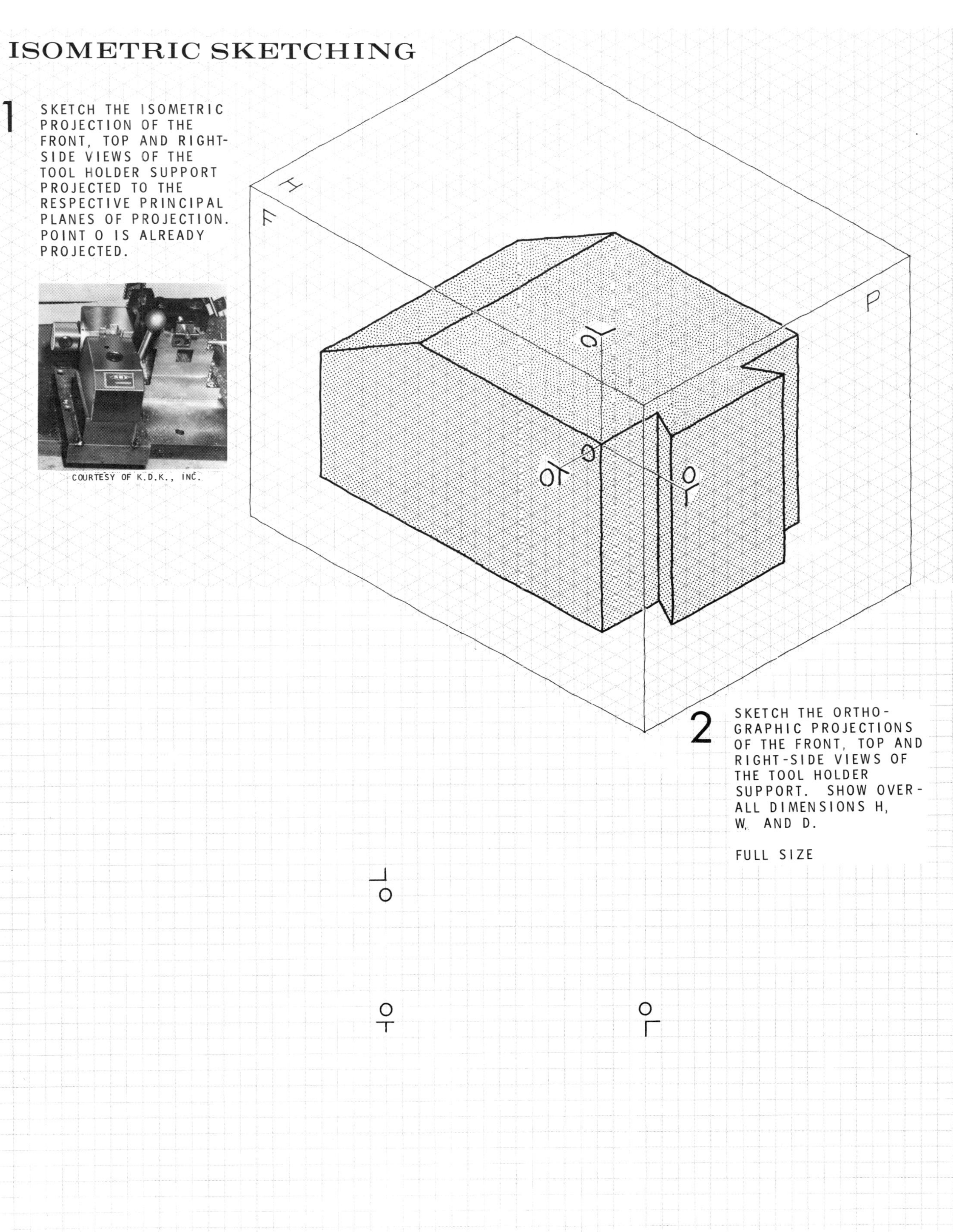


COURTESY OF K.D.K., INC.

**2** SKETCH THE ORTHOGRAPHIC PROJECTIONS OF THE FRONT, TOP AND RIGHT-SIDE VIEWS OF THE TOOL HOLDER SUPPORT. SHOW OVER-ALL DIMENSIONS H, W, AND D.

FULL SIZE

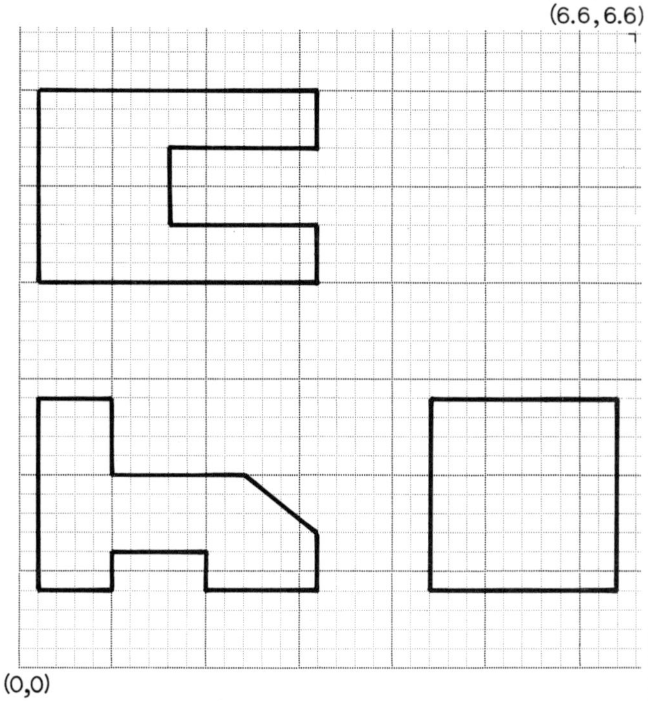

(6.6,6.6)

(0,0)

## ORTHOGRAPHIC VIEWS

Complete the three views of the part. Label the views and dimension them with D, W, and H.

# ORTHOGRAPHICS

1 DRAW THE MISSING VIEWS OF THE THREE-VIEW DRAWINGS.

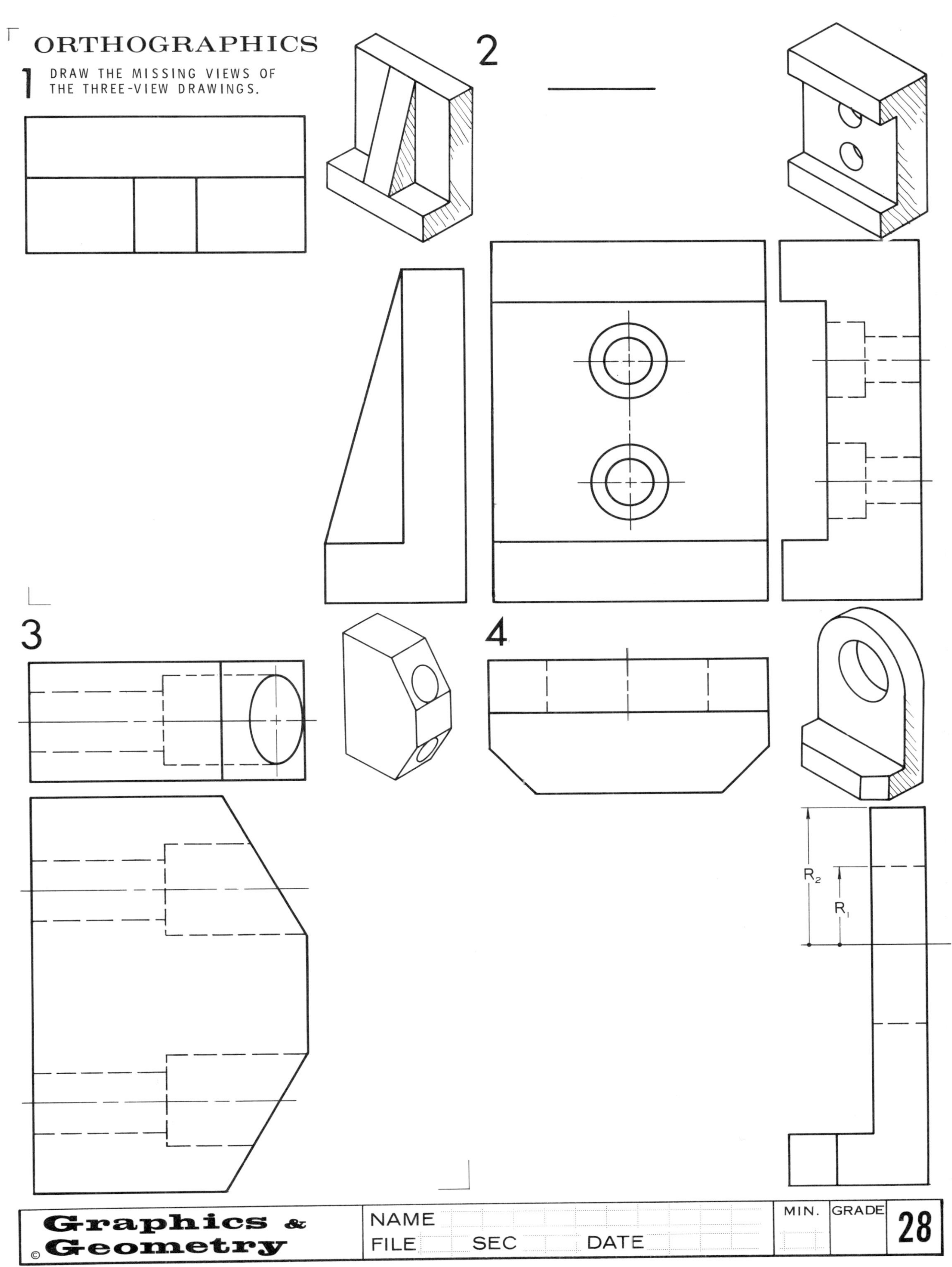

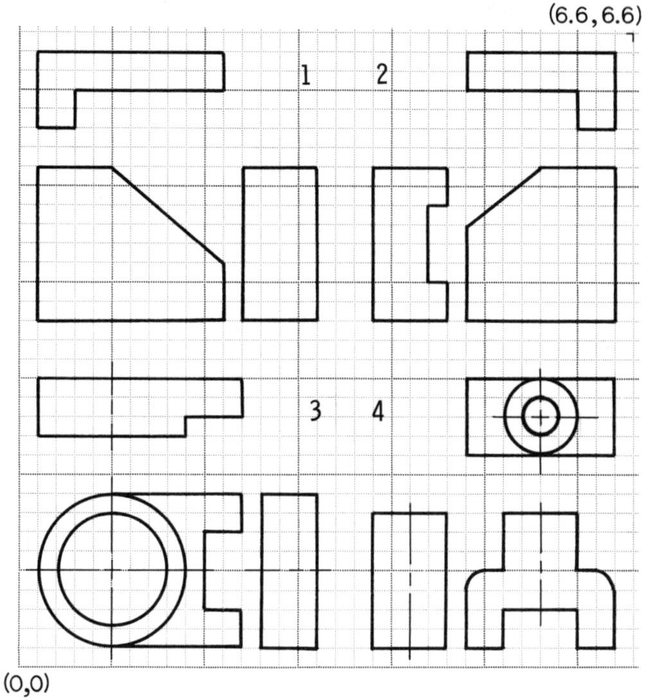

(6.6, 6.6)

1 2

3 4

(0,0)

ORTHOGRAPHIC PROJECTION.

Plot the views and supply the missing lines.

| COMPUTER GRAPHICS | Graphics for Engineers © | NAME<br>FILE SEC DATE | MIN | GRADE | 28 |
|---|---|---|---|---|---|

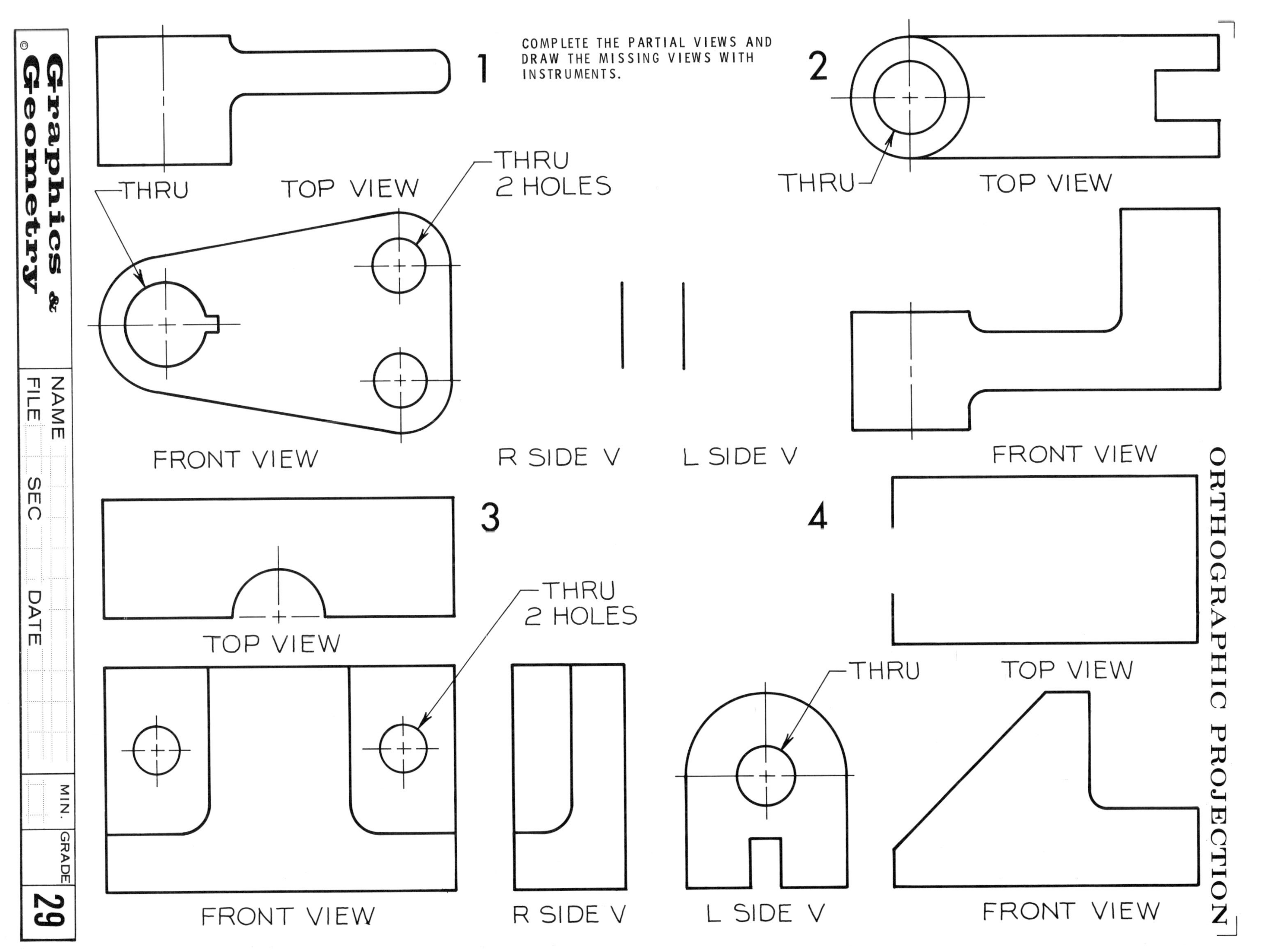
ORTHOGRAPHIC PROJECTION
COMPLETE THE PARTIAL VIEWS AND DRAW THE MISSING VIEWS WITH INSTRUMENTS.
1
THRU
TOP VIEW
THRU
2 HOLES
FRONT VIEW
R SIDE V
L SIDE V
2
THRU
TOP VIEW
FRONT VIEW
3
TOP VIEW
THRU
2 HOLES
FRONT VIEW
R SIDE V
4
THRU
TOP VIEW
L SIDE V
FRONT VIEW
Graphics & Geometry
NAME
FILE
SEC
DATE
MIN.
GRADE
29

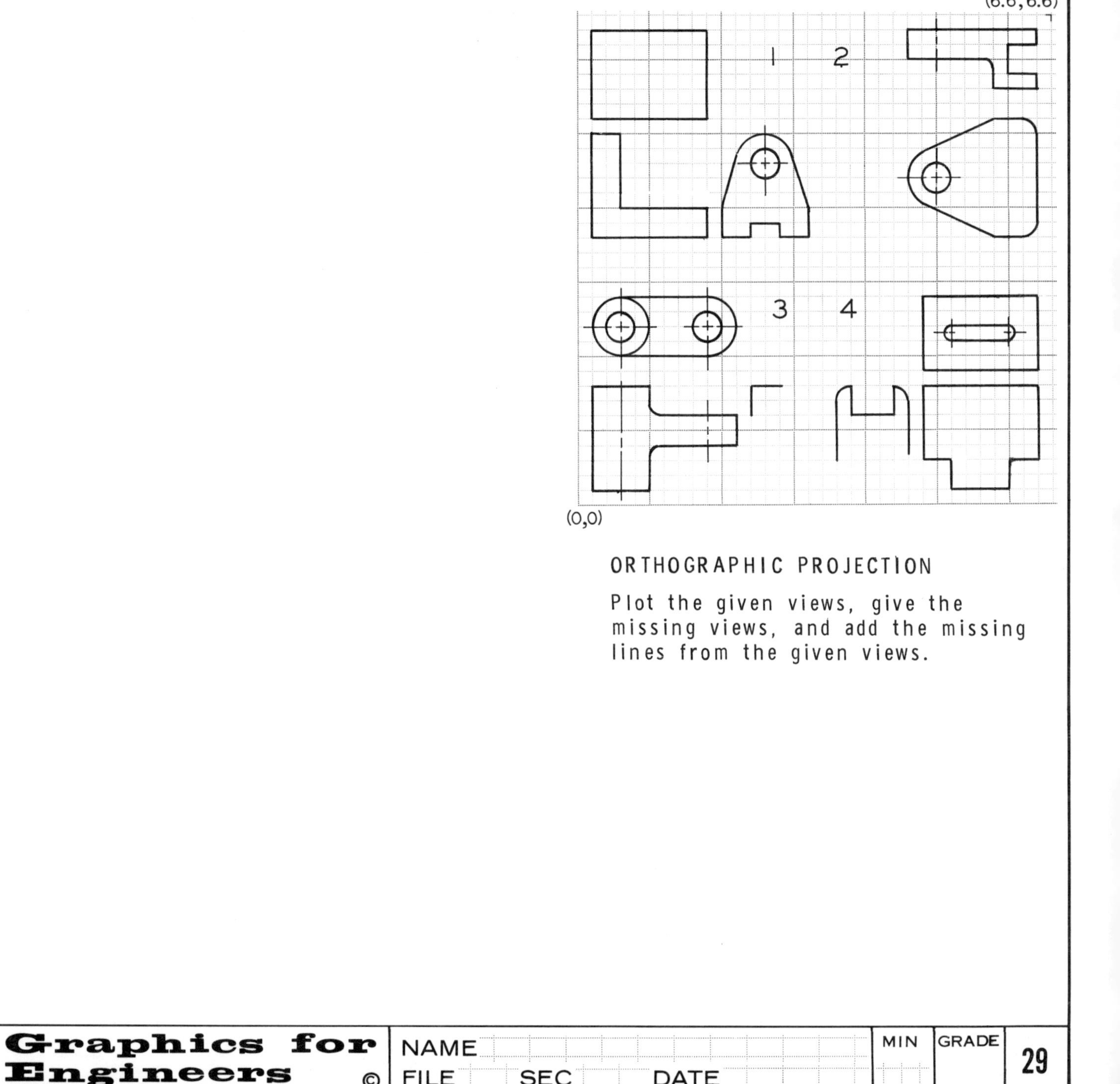


## ORTHOGRAPHIC PROJECTION

Plot the given views, give the missing views, and add the missing lines from the given views.

# ORTHOGRAPHICS

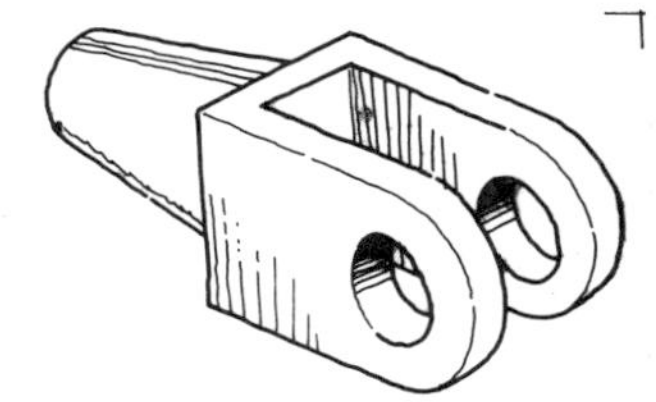

ENGINEERS FREQUENTLY MAKE FREEHAND SKETCHES TO CONVEY IDEAS RAPIDLY TO THEIR DRAFTSMEN WHO CONVERT THEM INTO INSTRUMENT DRAWINGS, CORRECTING ERRORS THAT MAY EXIST IN THE SKETCHES. ASSUME THAN AN ENGINEER HAS MADE THE SKETCHES BELOW, AND YOU ARE RESPONSIBLE FOR PREPARING INSTRUMENT DRAWINGS OF THE ORTHOGRAPHIC VIEWS. OMIT DIMENSIONS AND LOOK FOR ERRORS IN THE SKETCHES.

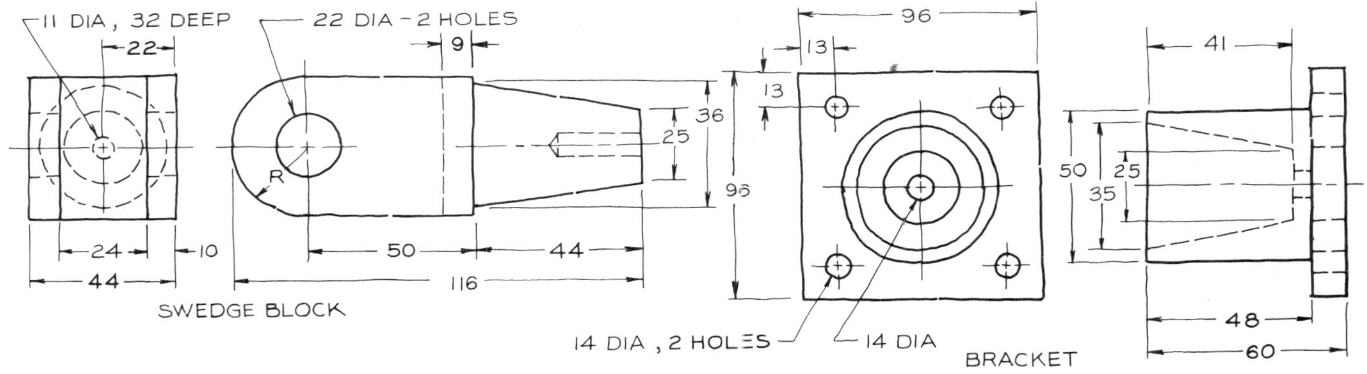


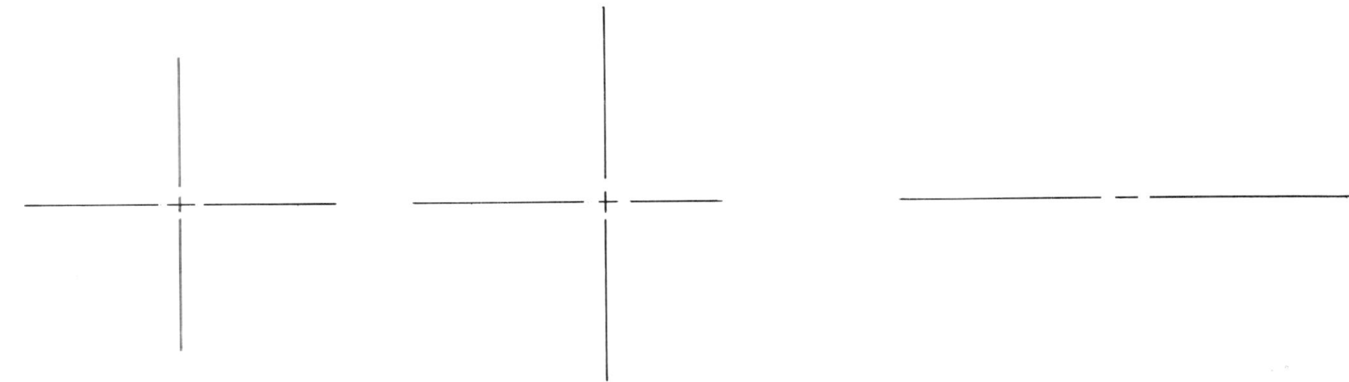


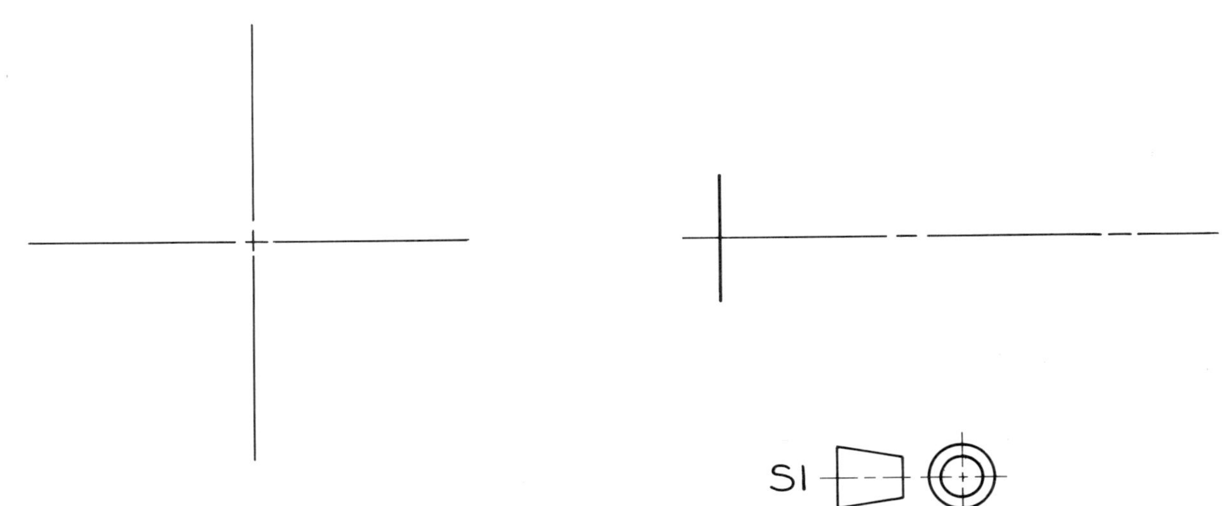


| Graphics & Geometry | NAME | MIN. | GRADE | 30 |
|---|---|---|---|---|
| | FILE SEC DATE | | | |

©

## ORTHOGRAPHICS

Follow the instructions on the opposite side of this sheet and plot the views of the parts: the Swedge Block and the Bracket.

| COMPUTER GRAPHICS | Graphics for Engineers © | NAME<br>FILE SEC DATE | MIN | GRADE | 30 |
|---|---|---|---|---|---|

## ORTHOGRAPHIC PROJECTION

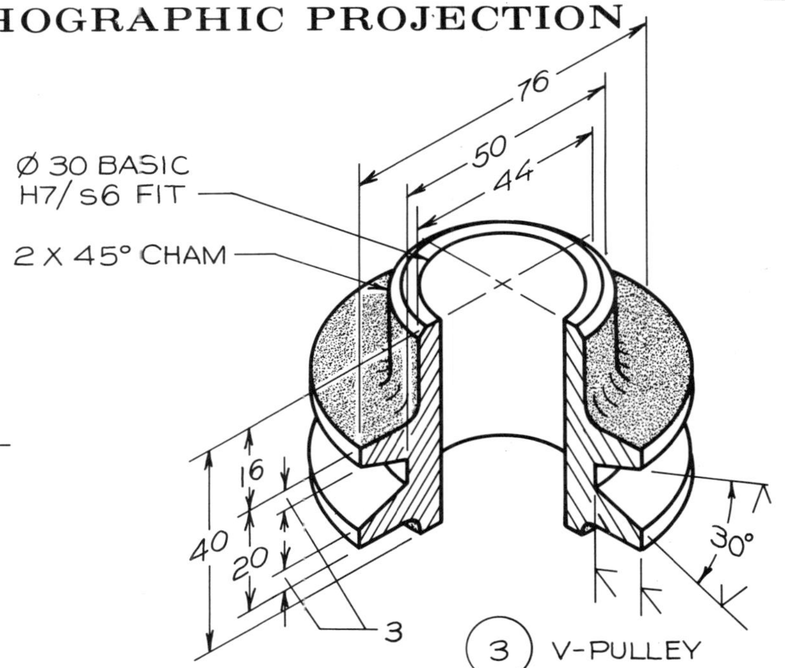

DRAW FULL-SIZE ORTHOGRAPHIC VIEWS OF THE V-PULLEY. THIS DRAWING WILL BE USED FOR THE WORKING DRAWINGS ON THE FOLDOUT SHEETS.

Ø 30 BASIC
H7/S6 FIT
2 X 45° CHAM
76
50
44
16
40
20
3
30°

(3) V-PULLEY
1020 STEEL
FILLETS & ROUNDS 3R

SI

## ORTHOGRAPHIC VIEWS

Follow the instructions on the opposite side of this sheet and make full-size orthographic views of the V-Pulley. Construct the views to be in a horizontal arrangement rather than vertical as shown.

| COMPUTER GRAPHICS | **Graphics for Engineers** © | NAME<br>FILE SEC DATE | MIN | GRADE | 31 |
|---|---|---|---|---|---|

CONVERT THE ORTHOGRAPHIC SKETCH OF THE IDLER BASE INTO THREE ORTHOGRAPHIC VIEWS DRAWN WITH INSTRUMENTS. THIS DRAWING WILL BE USED FOR THE WORKING DRAWING ON THE FOLDOUT SHEETS.

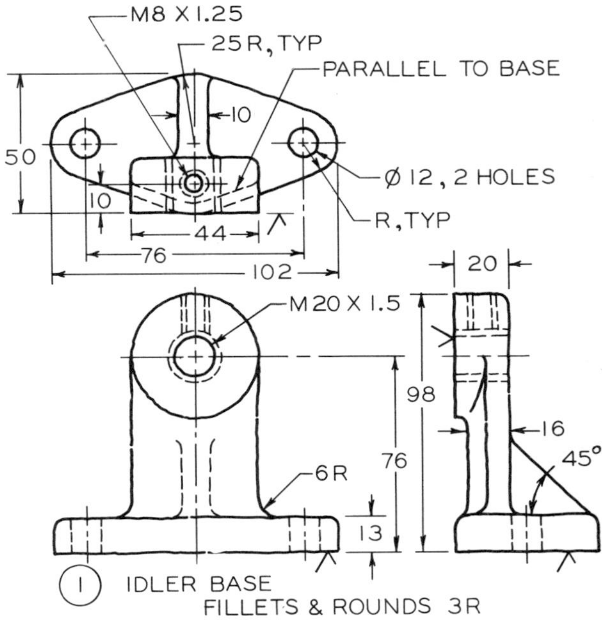


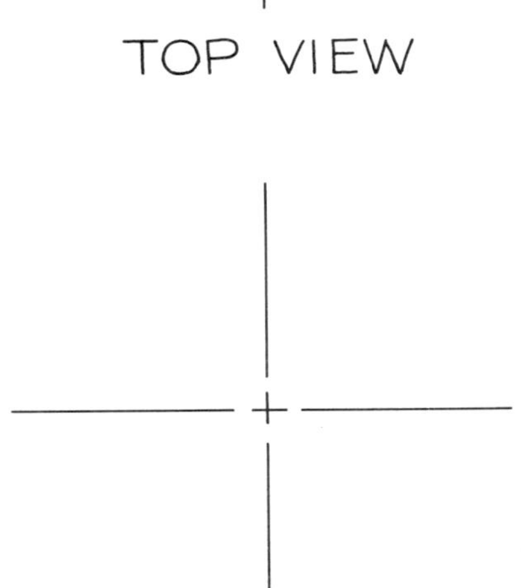

TOP VIEW

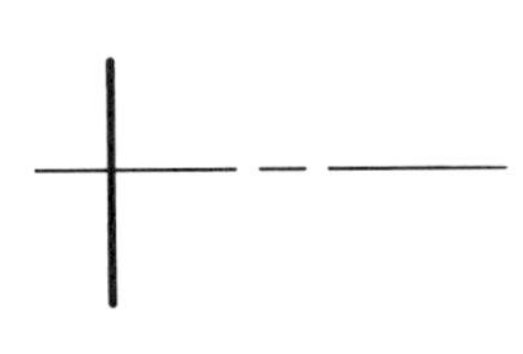

FRONT VIEW - FULL SIZE

R SIDE VIEW

(1) IDLER BASE SAE G2500-CI

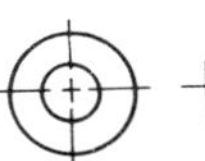

SI

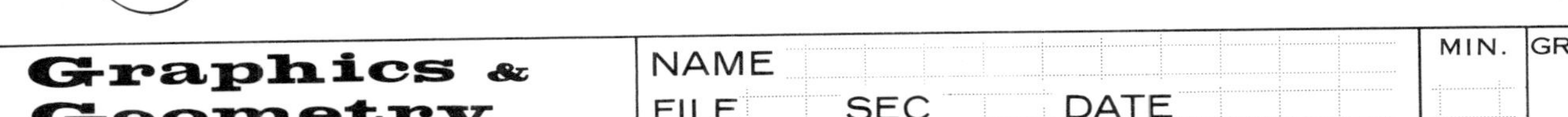

## ORTHOGRAPHIC PROJECTION

Plot the views of the Idler Base in accordance with the dimensions and specifications given on the opposite side of this sheet.

# AUXILIARY VIEWS

1

PROB 1 & 3: PROJECT THE VIEWS OF THE OBJECT ONTO THE PICTORIAL PROJECTION PLANES INCLUDING THE AUXILIARY VIEWS OF THE INCLINED SURFACES.

PROB 2 & 4: CONSTRUCT TRUE-SIZE AUXILIARY VIEWS OF THE INCLINED SURFACES.

2

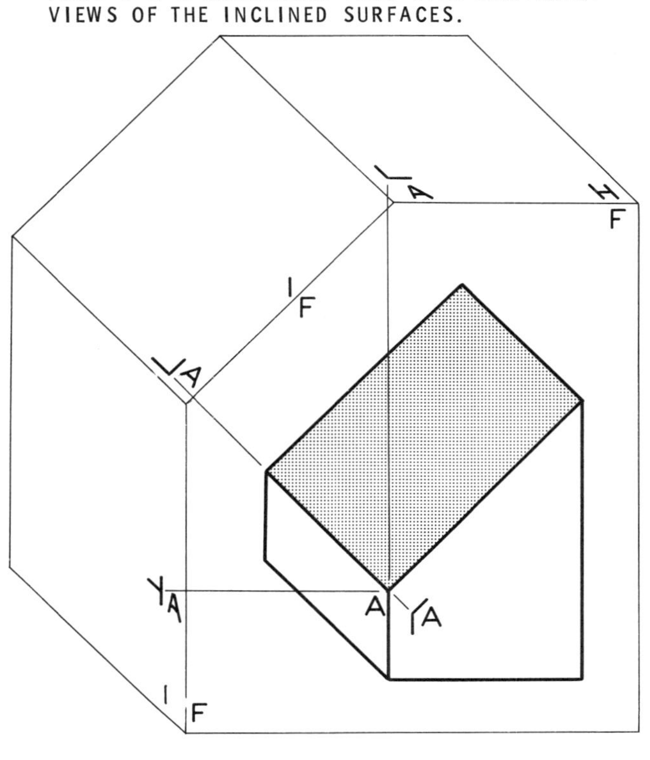


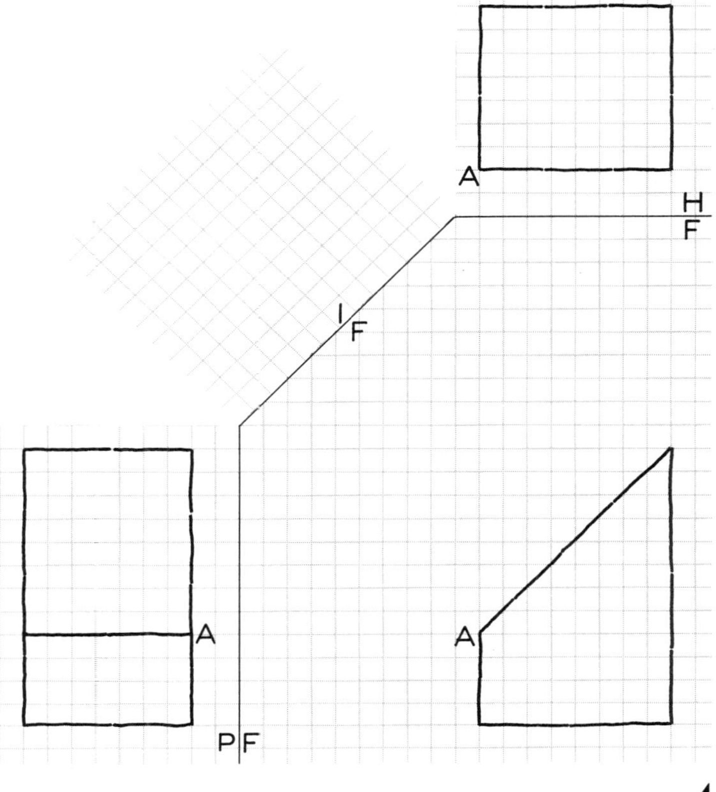


3

4

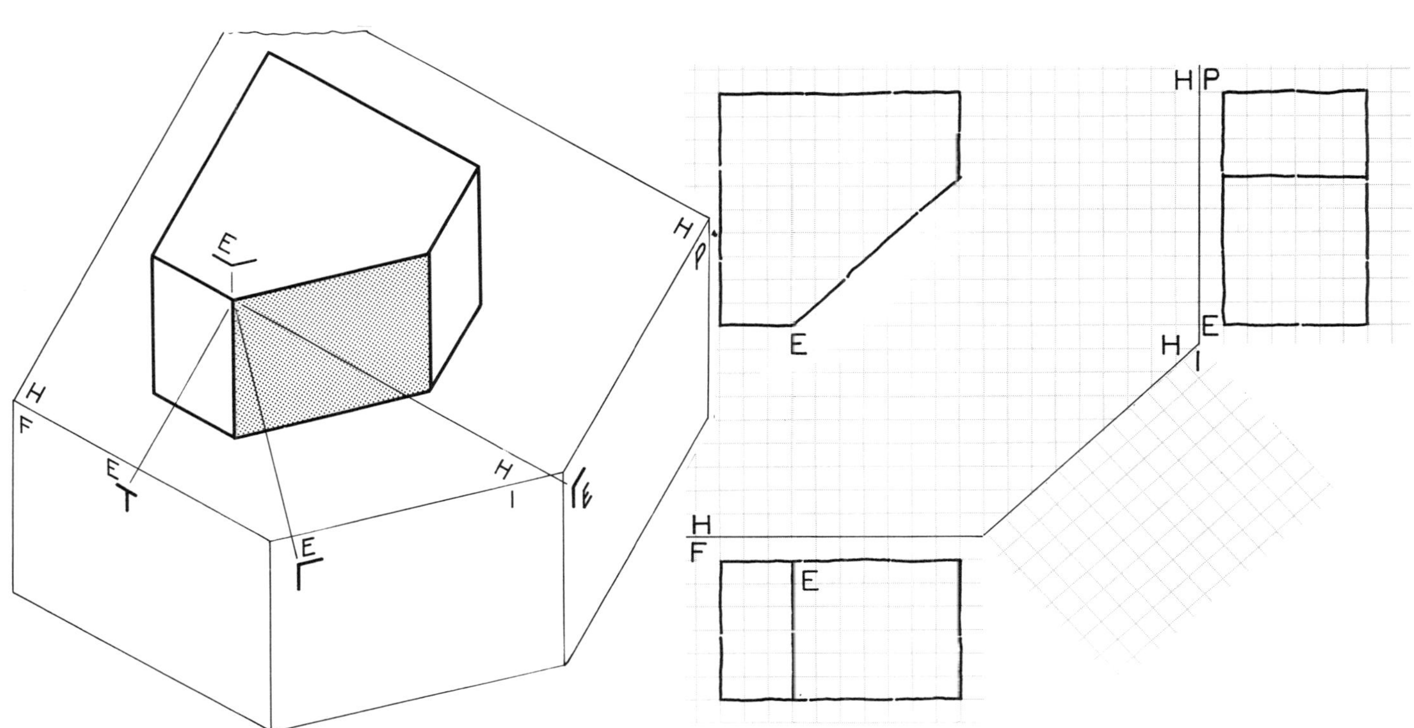

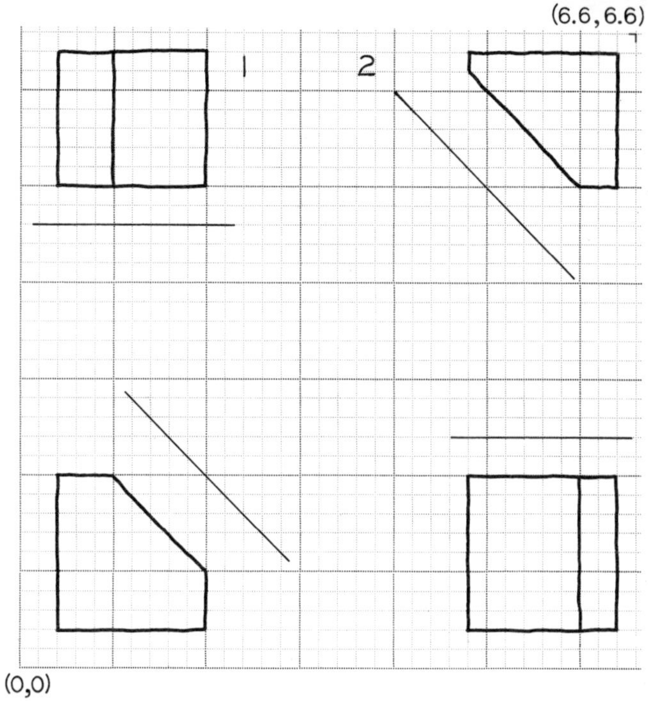

(6.6, 6.6)

1

2

(0,0)

## AUXILIARY VIEWS

Find the true-size views of the inclined surfaces by auxiliary views using the given reference lines.

DRAW TRUE-SIZE AUXILIARY VIEWS OF THE INCLINED SURFACES ONLY. NUMBER THE POINTS IN PROBLEM 1.

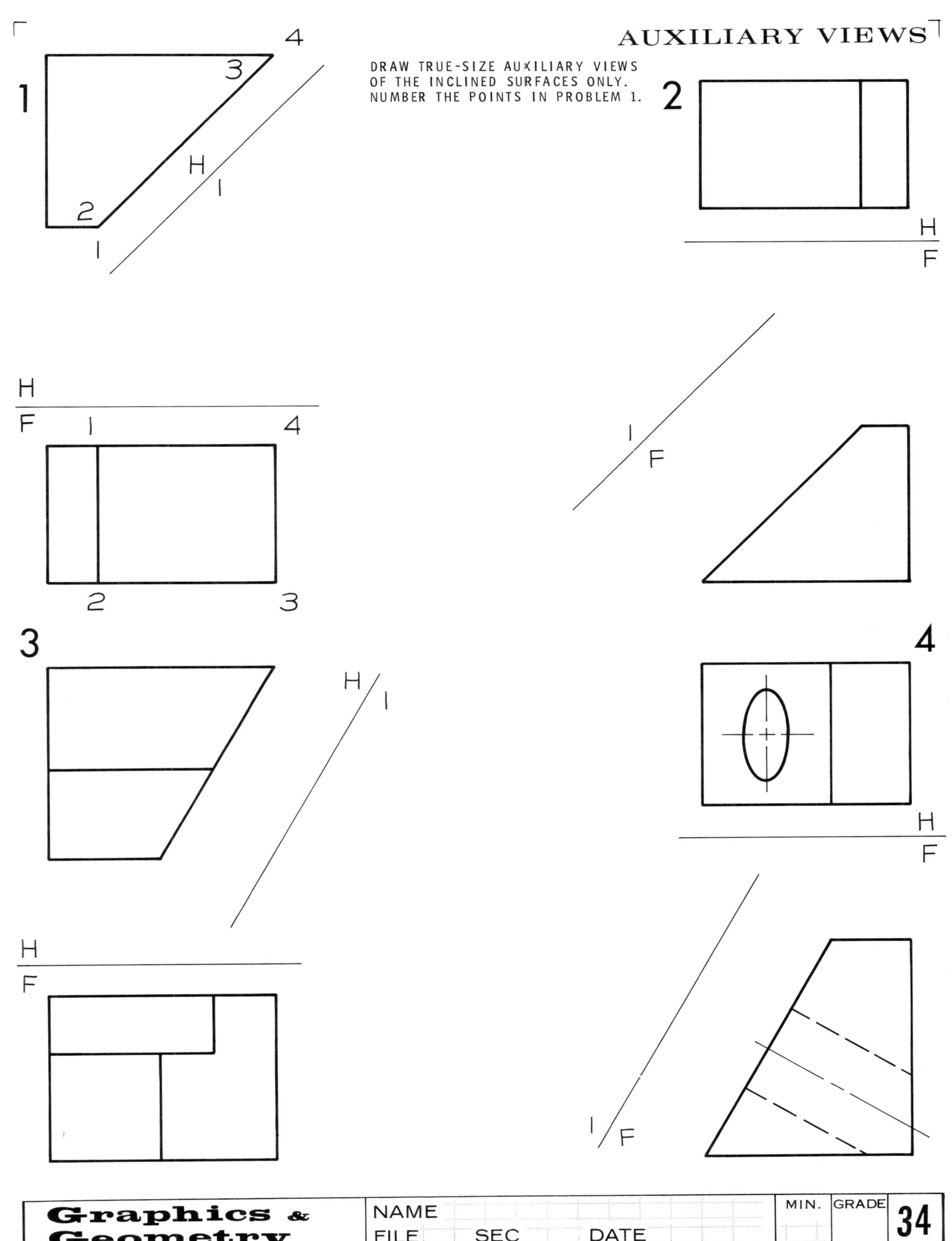


Graphics & Geometry

NAME

FILE SEC DATE

MIN. GRADE



©

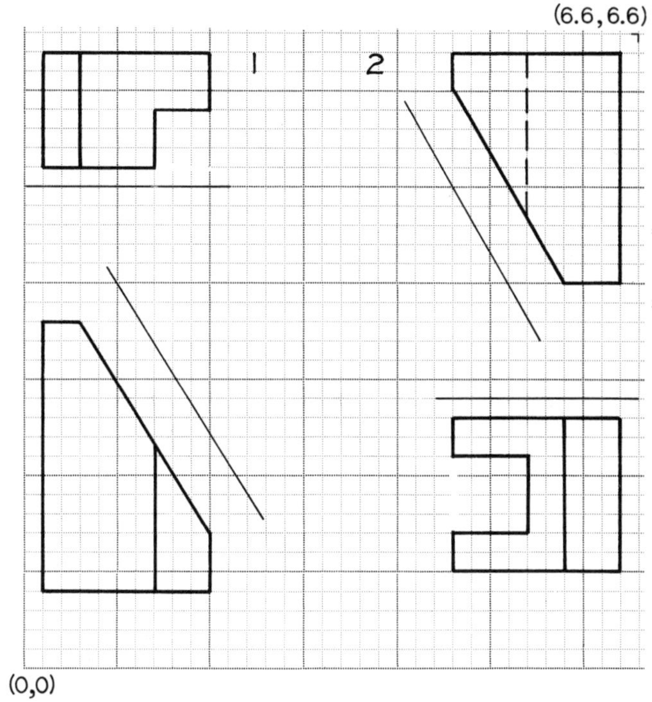

(6.6,6.6)

1

2

(0,0)

## AUXILIARY VIEWS

Find the true-size views of the inclined surfaces by auxiliary views using the given reference lines.



NAME

FILE SEC DATE

MIN GRADE

# AUXILIARY VIEWS

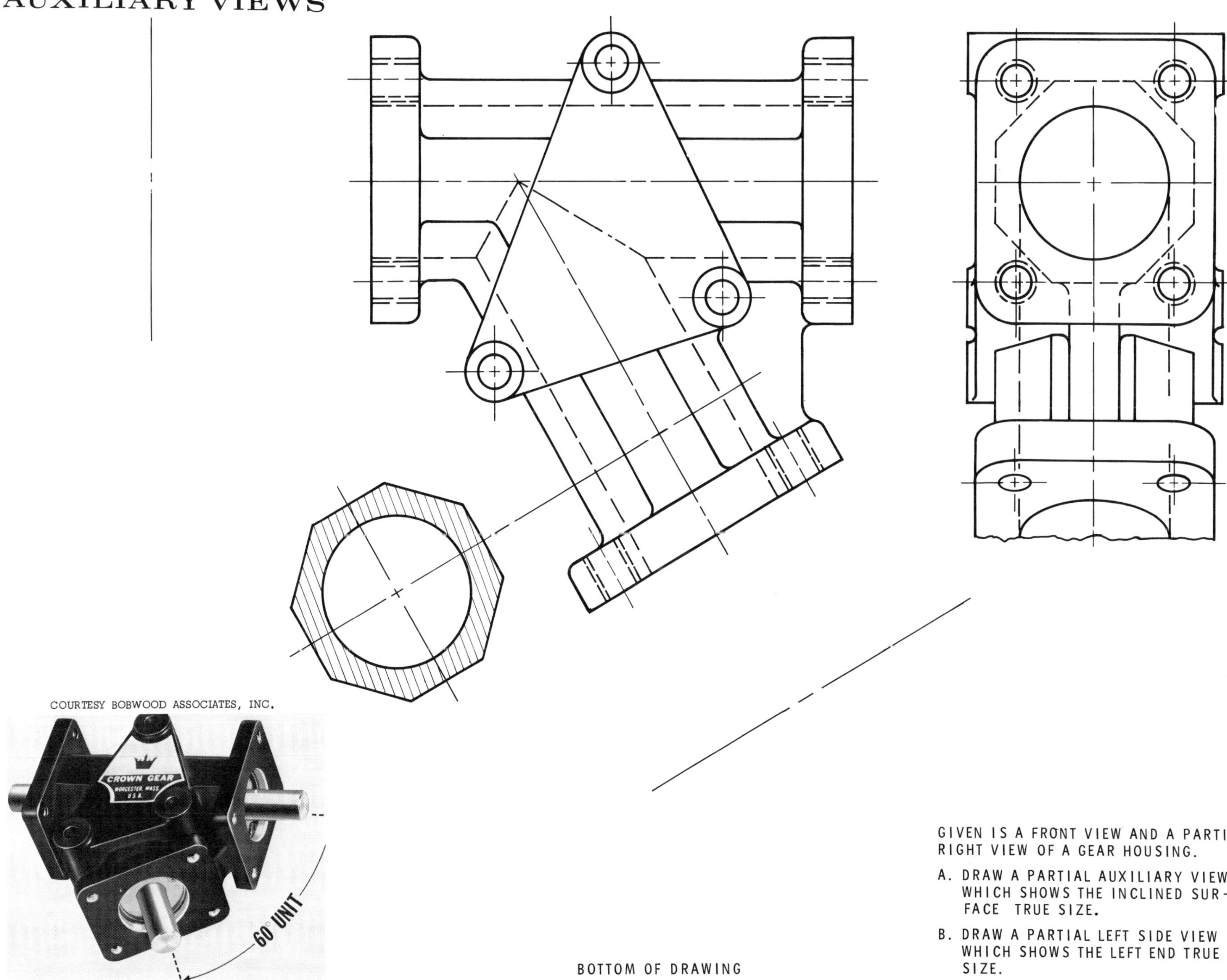


COURTESY BOBWOOD ASSOCIATES, INC.

BOTTOM OF DRAWING

GIVEN IS A FRONT VIEW AND A PARTIAL RIGHT VIEW OF A GEAR HOUSING.

A. DRAW A PARTIAL AUXILIARY VIEW WHICH SHOWS THE INCLINED SURFACE TRUE SIZE.

B. DRAW A PARTIAL LEFT SIDE VIEW WHICH SHOWS THE LEFT END TRUE SIZE.

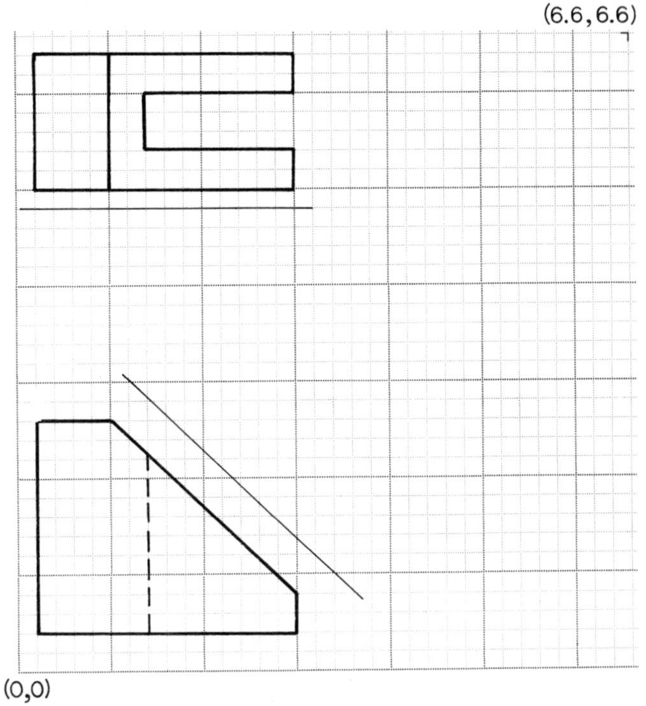

(6.6, 6.6)

(0,0)

## AUXILIARY VIEWS

Plot the views of the object and find the true-size view of the inclined surface by an auxiliary view.

| COMPUTER GRAPHICS | Graphics for Engineers © | NAME<br>FILE SEC DATE | MIN | GRADE | 35 |
|---|---|---|---|---|---|

# FULL SECTIONS

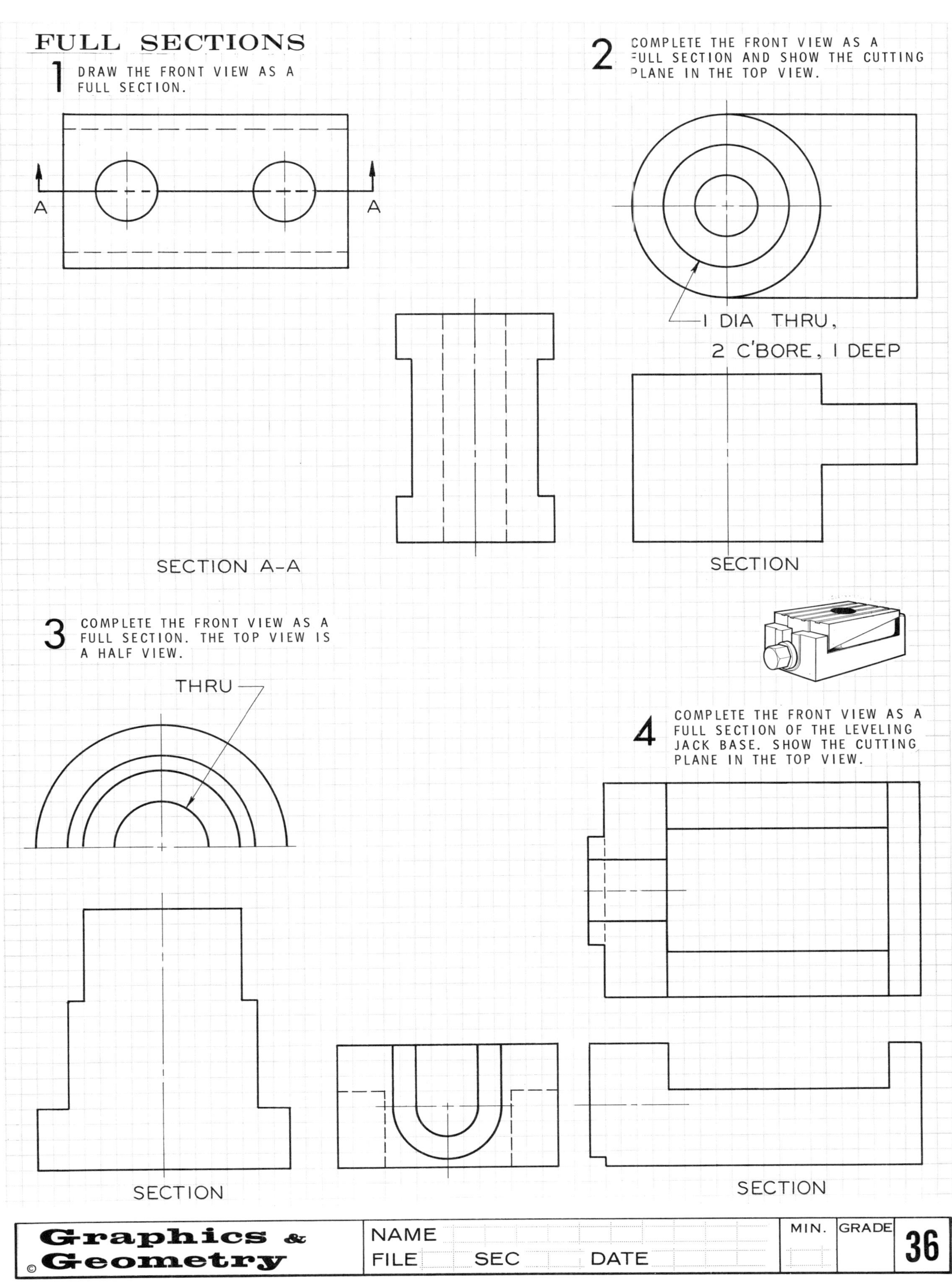

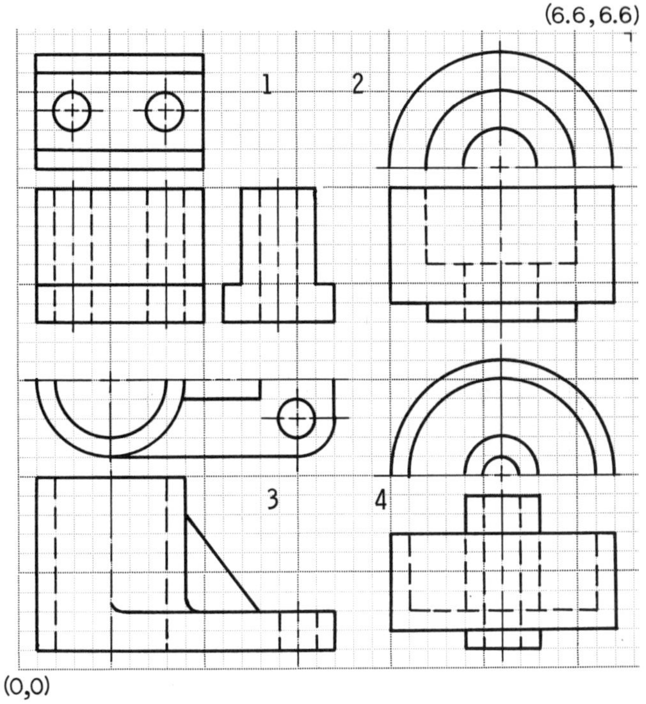

(6.6, 6.6)

1 2

3 4

(0,0)

## FULL SECTIONS

Complete the front views of the parts above as full sections. Hatch the areas that have been cut by the cutting plane. Omit unnecessary hidden lines.

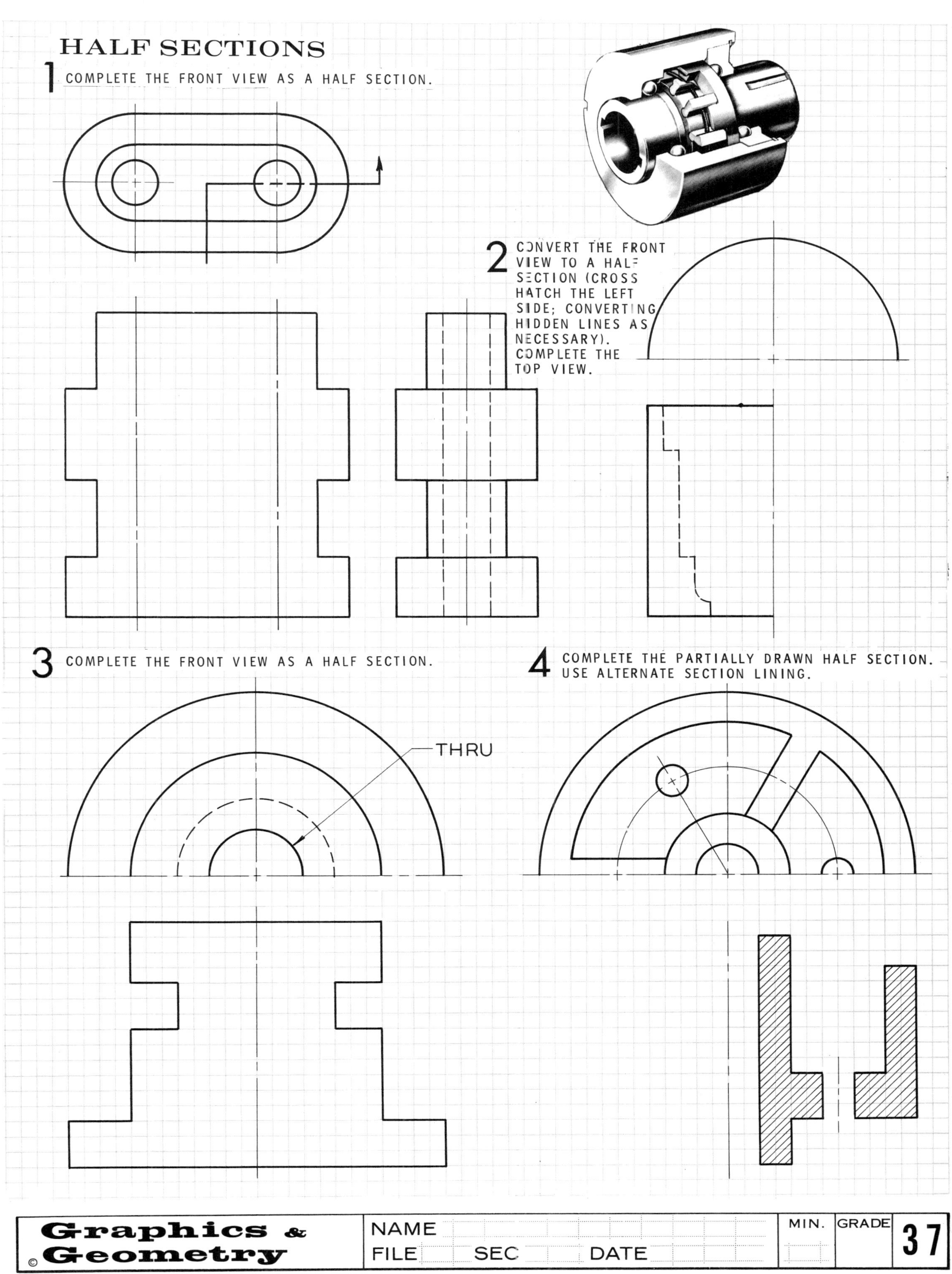
HALF SECTIONS
1 COMPLETE THE FRONT VIEW AS A HALF SECTION.
2 CONVERT THE FRONT VIEW TO A HALF SECTION (CROSS HATCH THE LEFT SIDE; CONVERTING HIDDEN LINES AS NECESSARY). COMPLETE THE TOP VIEW.
3 COMPLETE THE FRONT VIEW AS A HALF SECTION.
THRU
4 COMPLETE THE PARTIALLY DRAWN HALF SECTION. USE ALTERNATE SECTION LINING.
Graphics & Geometry
NAME
FILE
SEC
DATE
MIN.
GRADE
37

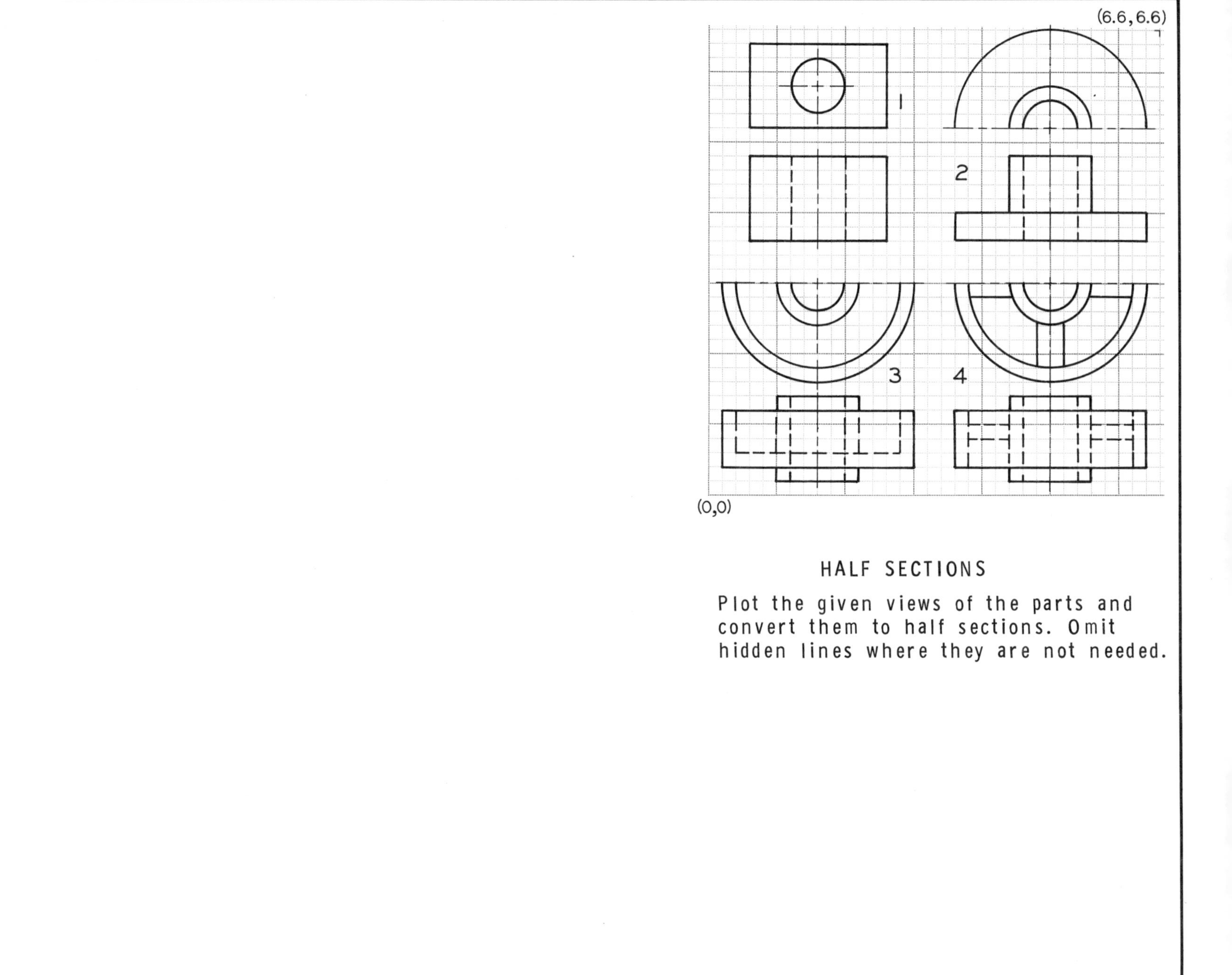


## HALF SECTIONS

Plot the given views of the parts and convert them to half sections. Omit hidden lines where they are not needed.

| COMPUTER GRAPHICS | Graphics for Engineers © | NAME<br>FILE SEC DATE | MIN | GRADE | 37 |
|---|---|---|---|---|---|

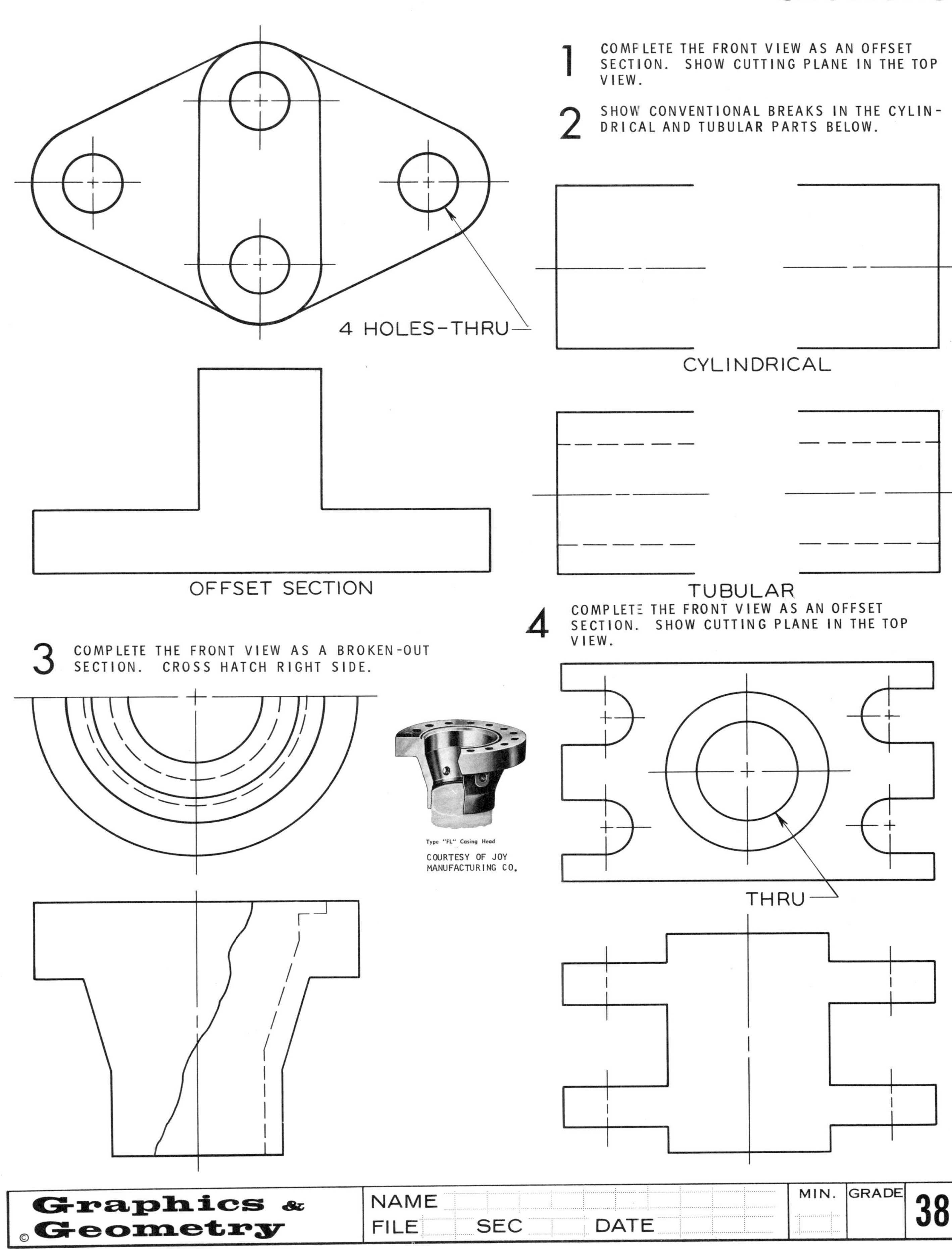
1
COMPLETE THE FRONT VIEW AS AN OFFSET SECTION. SHOW CUTTING PLANE IN THE TOP VIEW.
2
SHOW CONVENTIONAL BREAKS IN THE CYLINDRICAL AND TUBULAR PARTS BELOW.
4 HOLES-THRU
CYLINDRICAL
TUBULAR
OFFSET SECTION
4
COMPLETE THE FRONT VIEW AS AN OFFSET SECTION. SHOW CUTTING PLANE IN THE TOP VIEW.
3
COMPLETE THE FRONT VIEW AS A BROKEN-OUT SECTION. CROSS HATCH RIGHT SIDE.
Type "FL" Casing Head
COURTESY OF JOY MANUFACTURING CO.
THRU

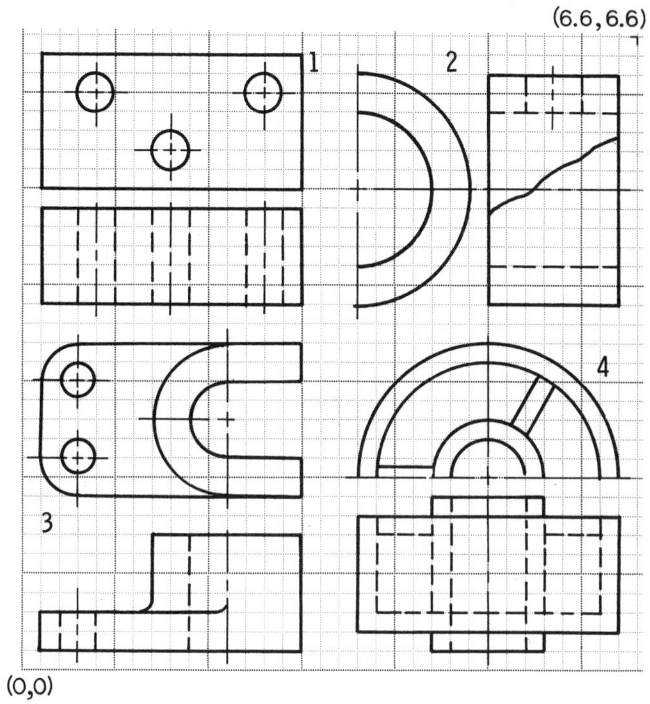

(6.6, 6.6)

1 2 3 4

(0,0)

## SECTIONS

Plot the views of the parts above and section them as follows:

1: Offset section

2: Broken-out section

3: Offset section

4: Full section

Omit hidden lines that are not needed.

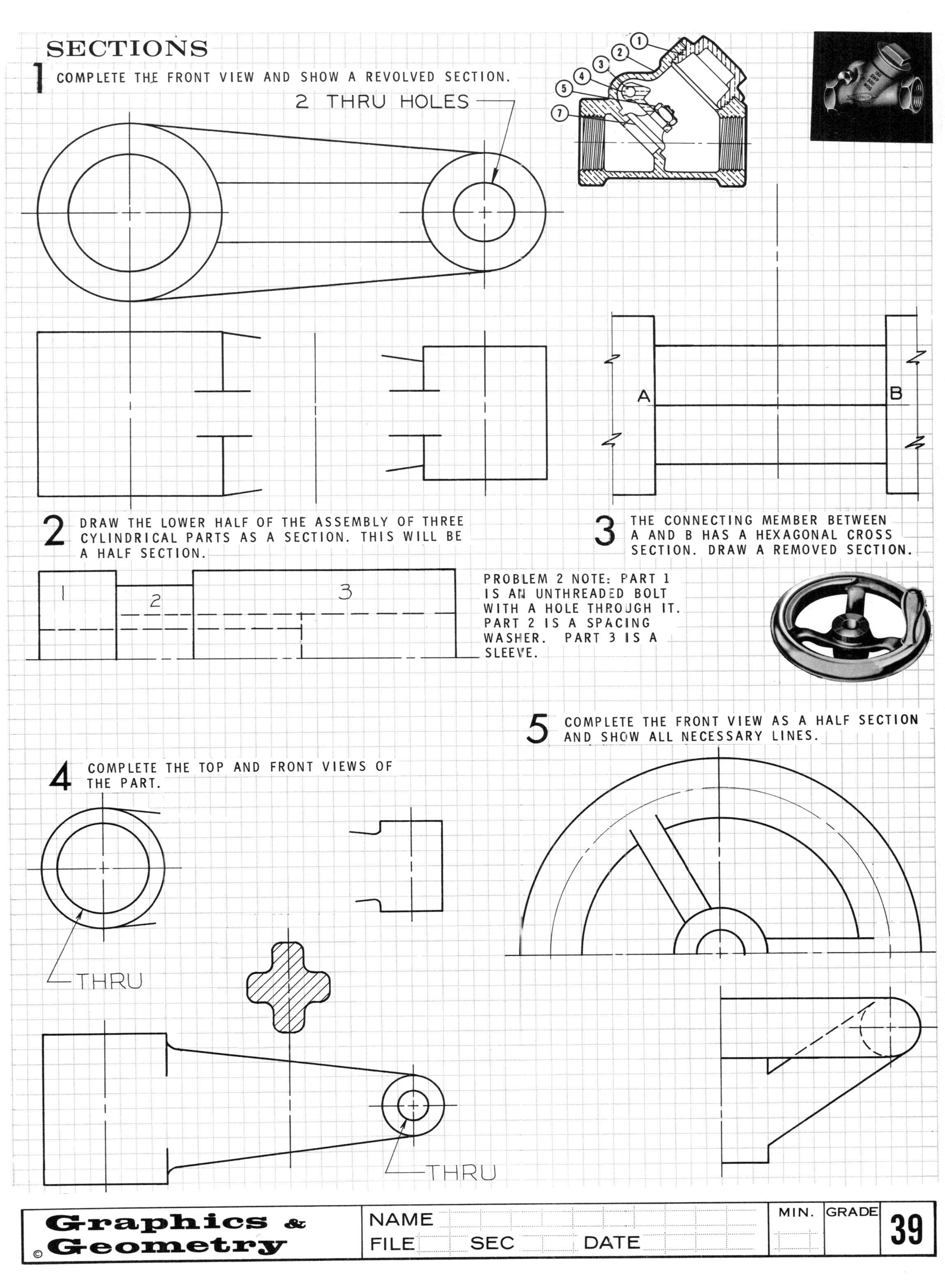

# SECTIONS

**1** COMPLETE THE FRONT VIEW AND SHOW A REVOLVED SECTION.



**2** DRAW THE LOWER HALF OF THE ASSEMBLY OF THREE CYLINDRICAL PARTS AS A SECTION. THIS WILL BE A HALF SECTION.

**3** THE CONNECTING MEMBER BETWEEN A AND B HAS A HEXAGONAL CROSS SECTION. DRAW A REMOVED SECTION.

PROBLEM 2 NOTE: PART 1 IS AN UNTHREADED BOLT WITH A HOLE THROUGH IT. PART 2 IS A SPACING WASHER. PART 3 IS A SLEEVE.

**5** COMPLETE THE FRONT VIEW AS A HALF SECTION AND SHOW ALL NECESSARY LINES.

**4** COMPLETE THE TOP AND FRONT VIEWS OF THE PART.

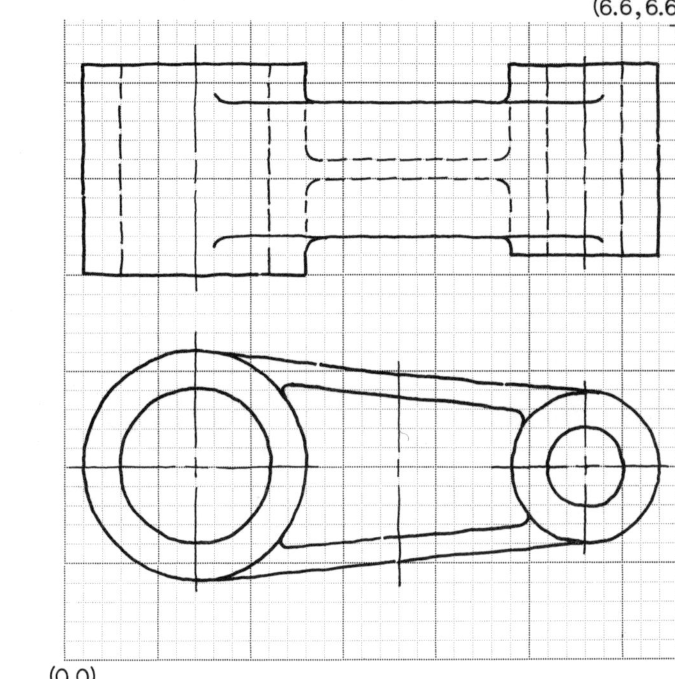


## SECTIONS

Plot the views of Column Arm and construct a revolved section in the front view about the axis that is given

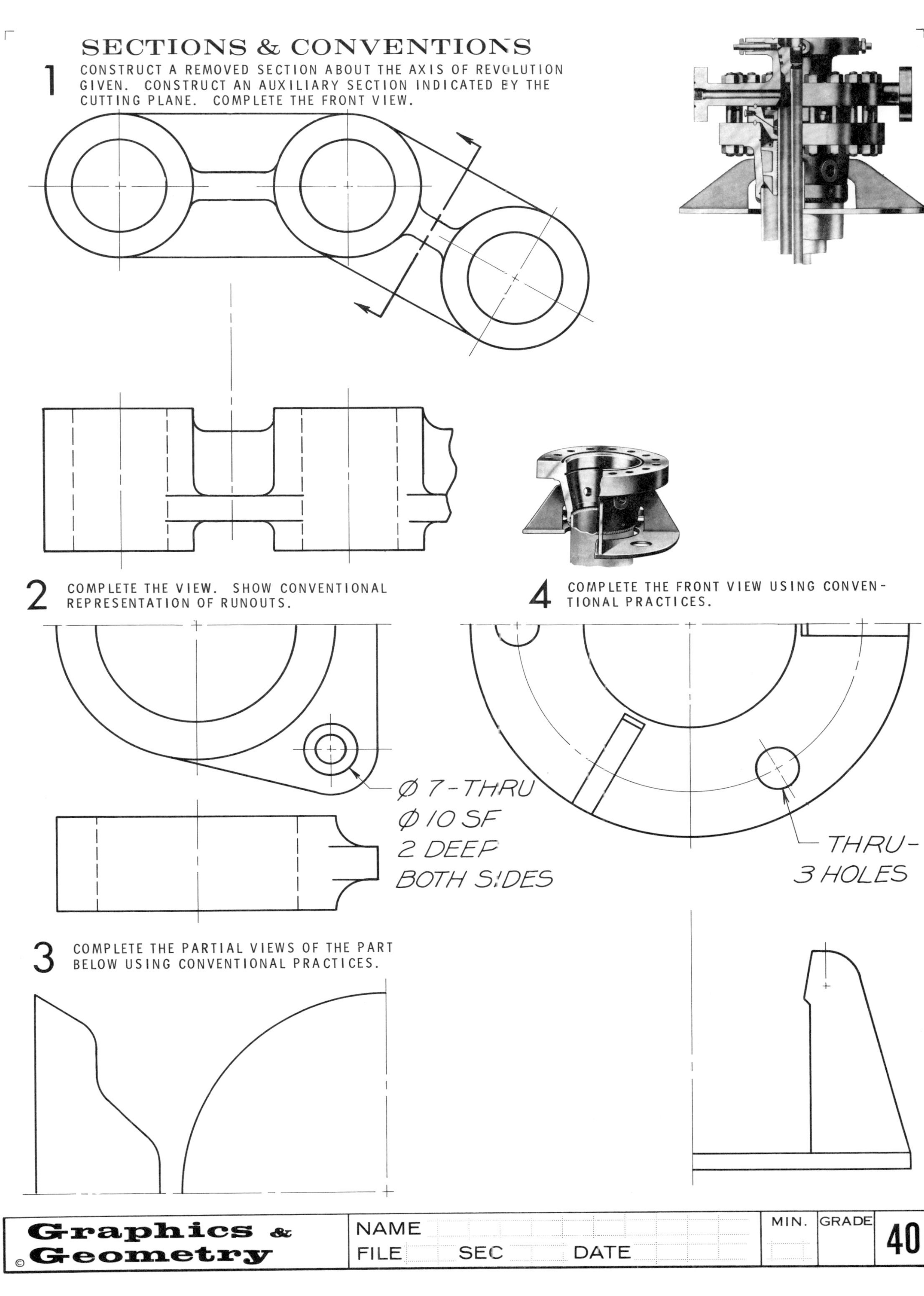
SECTIONS & CONVENTIONS
1
CONSTRUCT A REMOVED SECTION ABOUT THE AXIS OF REVOLUTION GIVEN. CONSTRUCT AN AUXILIARY SECTION INDICATED BY THE CUTTING PLANE. COMPLETE THE FRONT VIEW.
2
COMPLETE THE VIEW. SHOW CONVENTIONAL REPRESENTATION OF RUNOUTS.
Ø 7-THRU
Ø 10 SF
2 DEEP
BOTH SIDES
3
COMPLETE THE PARTIAL VIEWS OF THE PART BELOW USING CONVENTIONAL PRACTICES.
4
COMPLETE THE FRONT VIEW USING CONVENTIONAL PRACTICES.
THRU-
3 HOLES
Graphics & Geometry
NAME
FILE
SEC
DATE
MIN.
GRADE
40

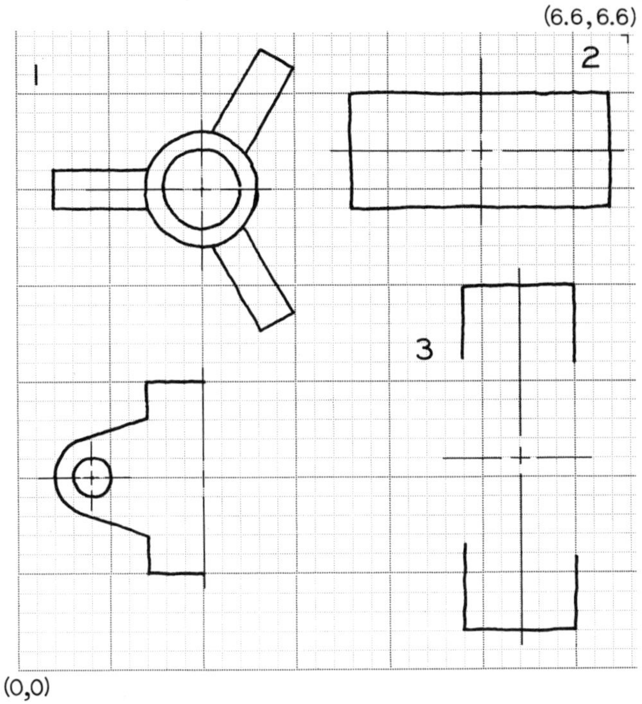

(6.6,6.6)

1

2

3

(0,0)

## CONVENTIONAL PRACTICES

Plot and complete the problems in the following manner:

1. Complete the views of the symmetrical part as a full section.

   Option 2: Complete the views of the part as conventional view.

2. Show a revolved section at the midpoint of the tubular part without conventional breaks.

3. Show a revolved section at the midpoint of the cast iron part that is square in section and show conventional breaks.

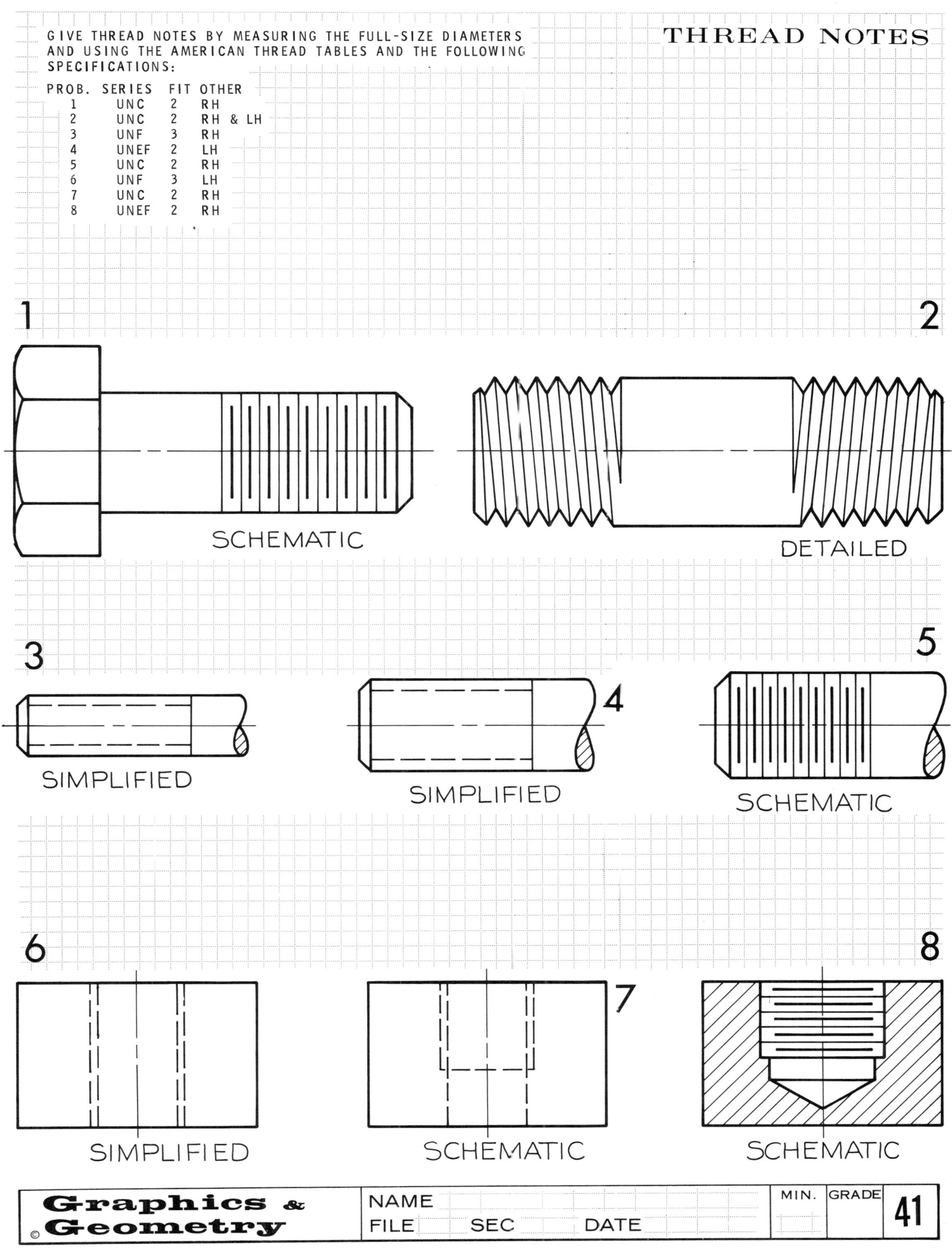
THREAD NOTES
GIVE THREAD NOTES BY MEASURING THE FULL-SIZE DIAMETERS AND USING THE AMERICAN THREAD TABLES AND THE FOLLOWING SPECIFICATIONS:
PROB. SERIES FIT OTHER
1 UNC 2 RH
2 UNC 2 RH & LH
3 UNF 3 RH
4 UNEF 2 LH
5 UNC 2 RH
6 UNF 3 LH
7 UNC 2 RH
8 UNEF 2 RH
1
SCHEMATIC
2
DETAILED
3
SIMPLIFIED
4
SIMPLIFIED
5
SCHEMATIC
6
SIMPLIFIED
7
SCHEMATIC
8
SCHEMATIC
Graphics & Geometry
©
NAME
FILE
SEC
DATE
MIN.
GRADE
41

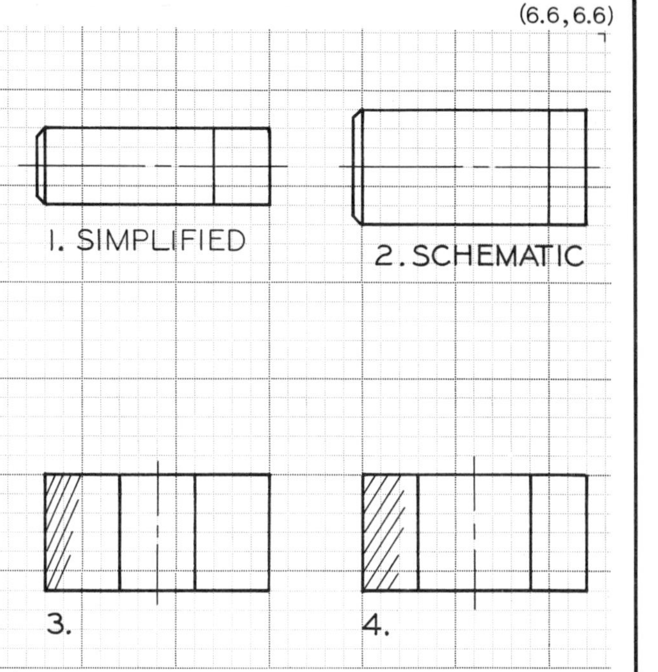


THREADED PARTS

Plot the threaded parts and show their thread symbols in accordance with the following specifications:

1. Simplified, .75-10 UNC-2A
2. Schematic, 1.25-12 UNF-2A
3. Simplified, .75-16 UNF-2B
4. Schematic, 1.25-7 UNC-2B

# THREADED PARTS

PROBS 1-4: DRAW THREAD REPRESENTATIONS AND NOTES FOR THE FULL-SIZE PARTS. USE IMPERIAL UNITS. CALCULATE THE THREAD LENGTHS. EACH IS UNC WITH A CLASS 2 FIT. 1 & 2: SIMPLIFIED SYMBOLS. 3 & 4: SCHEMATIC SYMBOLS. IDENTIFY THE HEADS OF EACH.

PROB 5: COMPLETE THE PARTIALLY DRAWN VIEW OF THE NUT.

PROB 6: COMPLETE THE VIEWS OF THE HEX HD BOLT DRAWN ACROSS CORNERS. CALCULATE THREAD LENGTH AND DRAW DETAILED THREAD SYMBOLS.

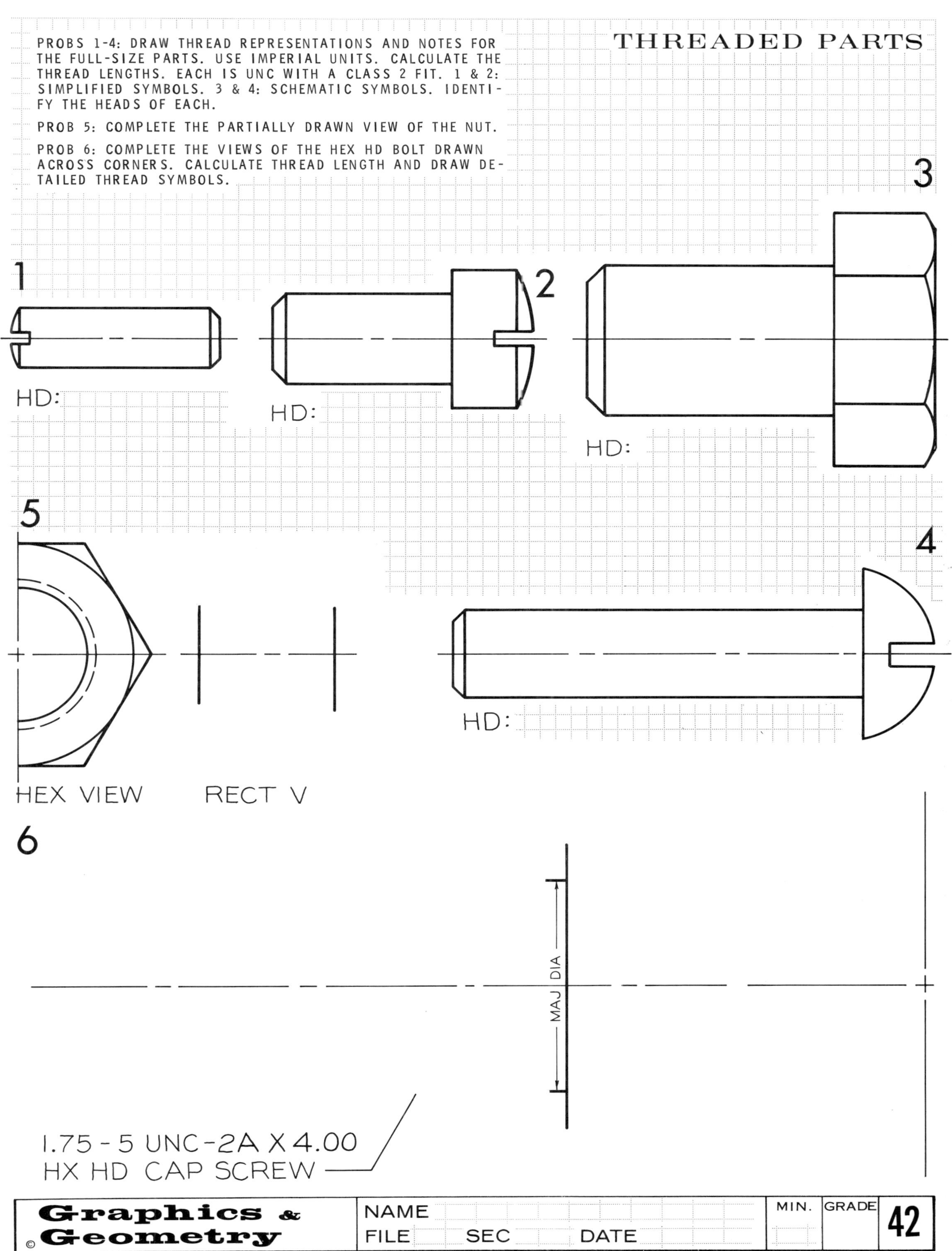


| Graphics & Geometry | NAME<br>FILE SEC DATE | MIN. | GRADE | 42 |
|---|---|---|---|---|

©

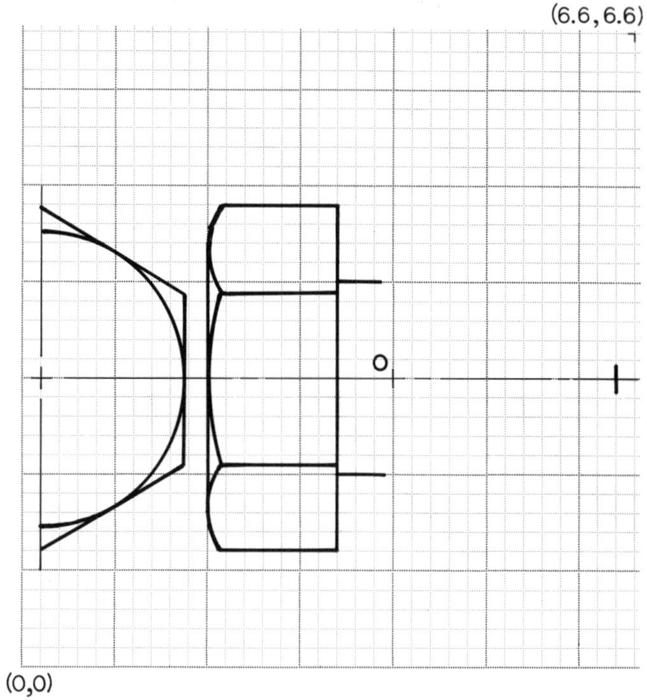

(6.6, 6.6)

O

(0,0)

DETAILED THREADS

Plot the hexagon head bolt, draw detailed threads from the end to point O, and give a thread note for the 2.00 DIA fine thread with a class 2 fit.

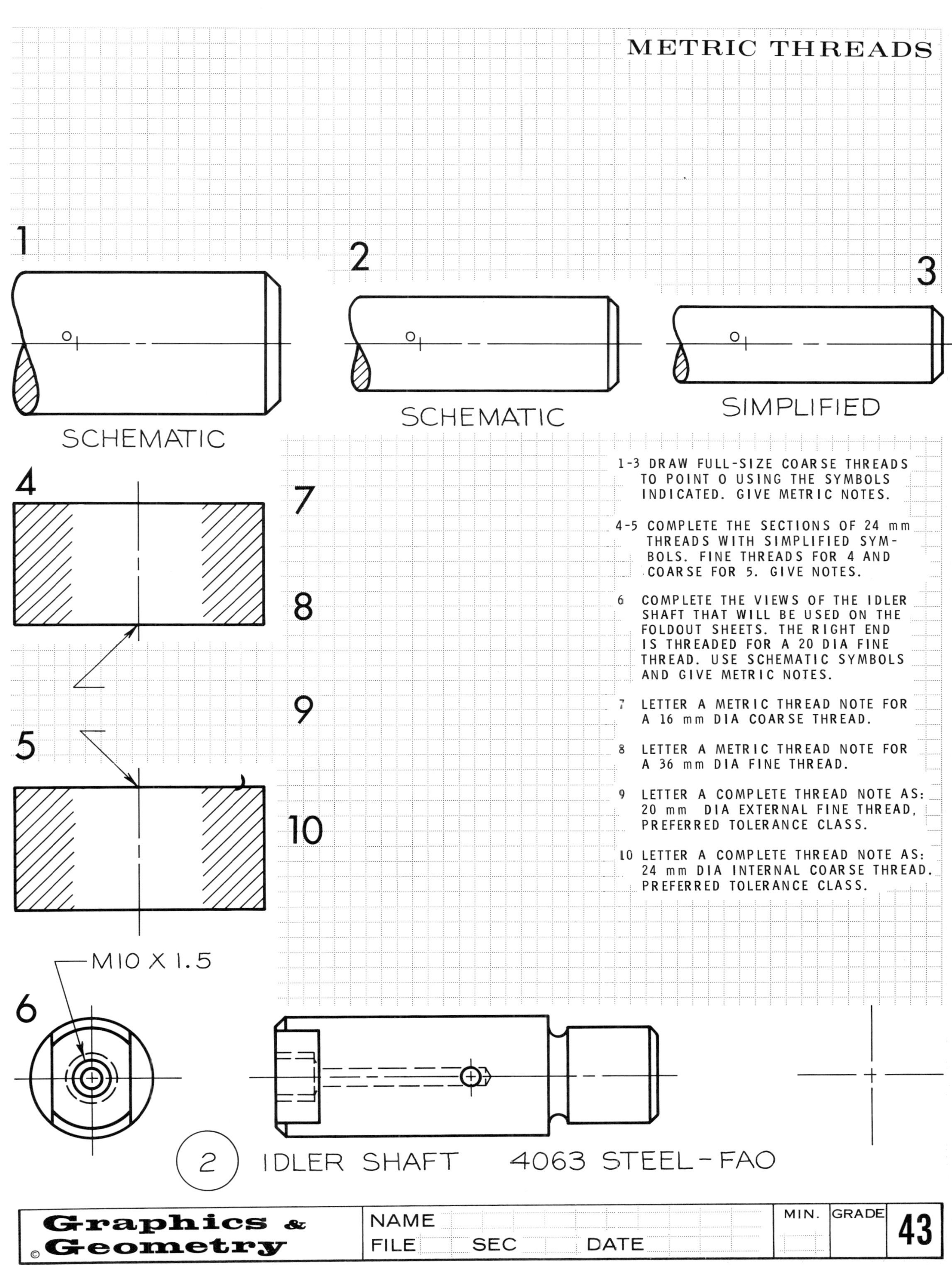


1-3 DRAW FULL-SIZE COARSE THREADS TO POINT O USING THE SYMBOLS INDICATED. GIVE METRIC NOTES.

4-5 COMPLETE THE SECTIONS OF 24 mm THREADS WITH SIMPLIFIED SYMBOLS. FINE THREADS FOR 4 AND COARSE FOR 5. GIVE NOTES.

6 COMPLETE THE VIEWS OF THE IDLER SHAFT THAT WILL BE USED ON THE FOLDOUT SHEETS. THE RIGHT END IS THREADED FOR A 20 DIA FINE THREAD. USE SCHEMATIC SYMBOLS AND GIVE METRIC NOTES.

7 LETTER A METRIC THREAD NOTE FOR A 16 mm DIA COARSE THREAD.

8 LETTER A METRIC THREAD NOTE FOR A 36 mm DIA FINE THREAD.

9 LETTER A COMPLETE THREAD NOTE AS: 20 mm DIA EXTERNAL FINE THREAD, PREFERRED TOLERANCE CLASS.

10 LETTER A COMPLETE THREAD NOTE AS: 24 mm DIA INTERNAL COARSE THREAD. PREFERRED TOLERANCE CLASS.

| Graphics & Geometry © | NAME<br>FILE SEC DATE | MIN. | GRADE | 43 |
|---|---|---|---|---|

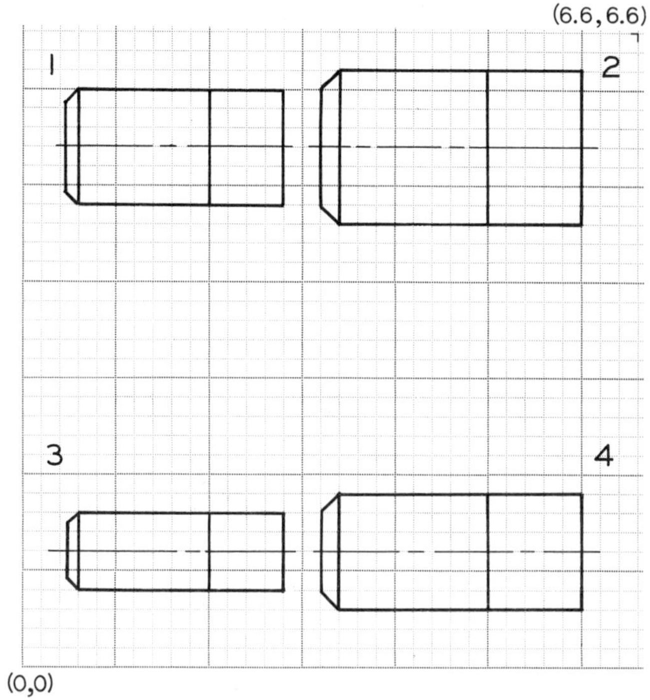

(6.6,6.6)

1

2

3

4

(0,0)

## METRIC THREADS

Plot the threaded parts above, show thread symbols and give their notes in accordance with the specifications below:

1. Schematic, Maj Dia = 30, Coarse
2. Schematic, Maj Dia = 42, Fine
3. Simplified, Maj Dia = 20, Coarse
4. Simplified, Maj Dia = 30, Fine

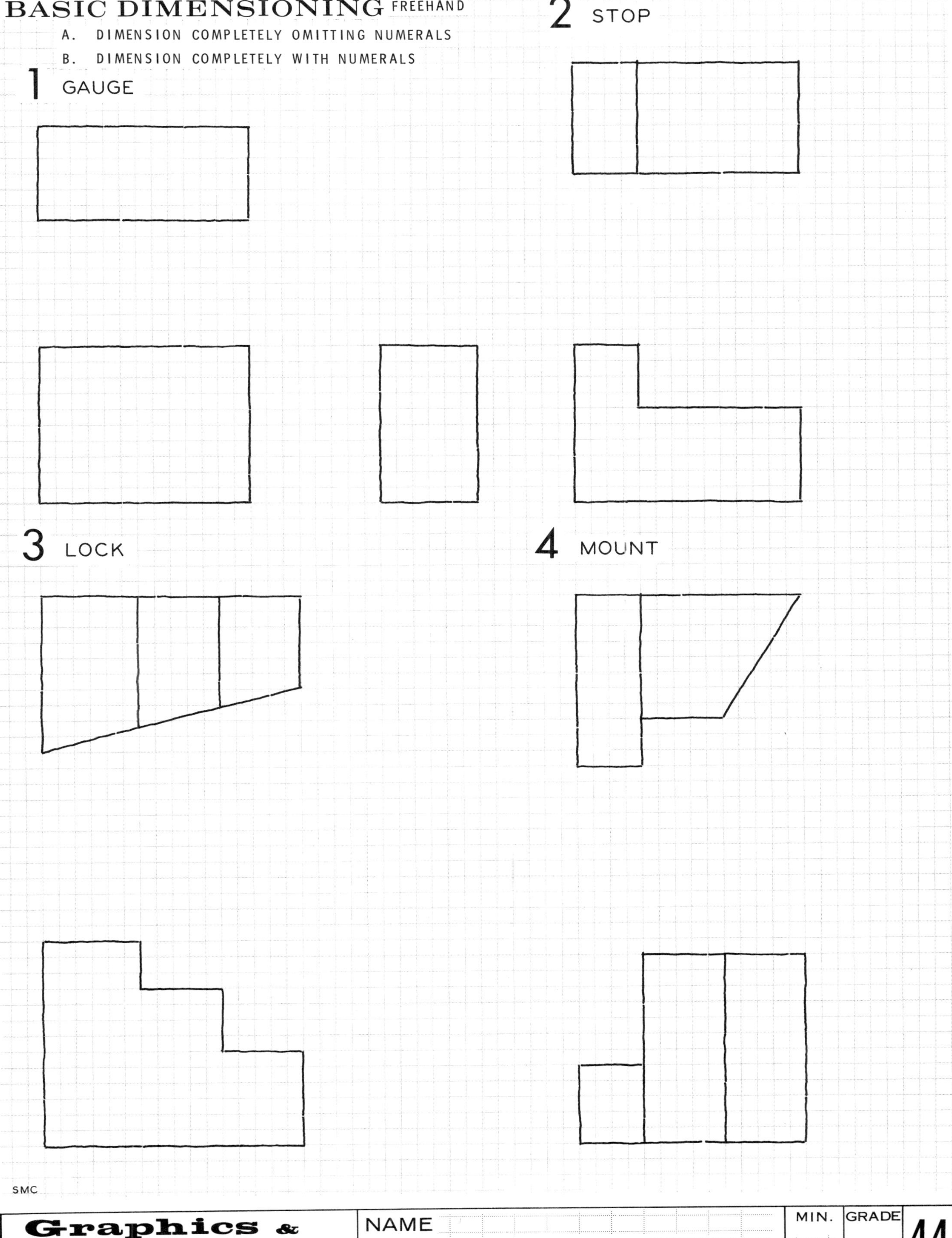


| Graphics & Geometry © | NAME<br>FILE SEC DATE | MIN. | GRADE | 44 |
|---|---|---|---|---|

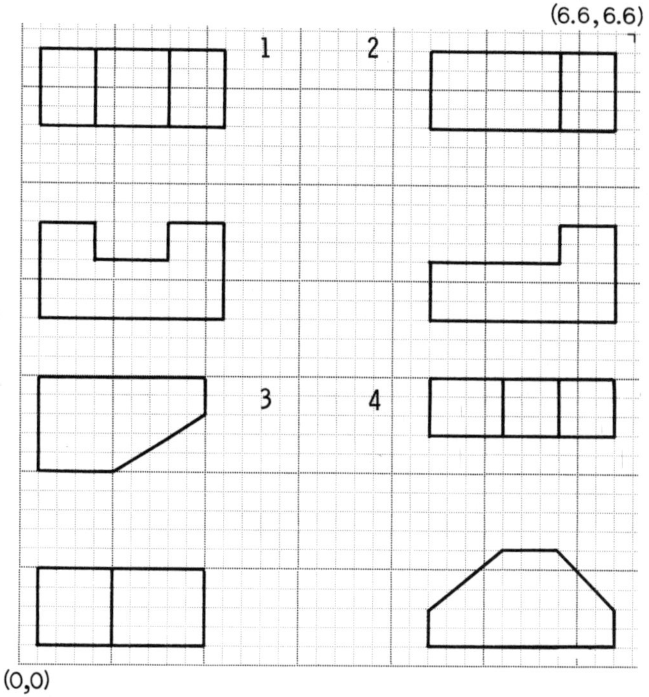

(6.6, 6.6)

1 2

3 4

(0,0)

## DIMENSIONING

Plot and dimension the views of the four parts. Use dimensioning text that is 0.07" high. Select other variable sizes that will fit in the available space. SCALE: FULL SIZE

# BASIC DIMENSIONING

DIMENSION FREEHAND. FOLLOW INSTRUCTIONS A OR B AS ASSIGNED. COUNT THE 1/8" GRID TO DETERMINE DIMENSIONS. SCALE: FULL SIZE.

A. DIMENSION COMPLETELY WITHOUT NUMERALS.

B. DIMENSION COMPLETELY USING NUMERALS.

1 SPACER

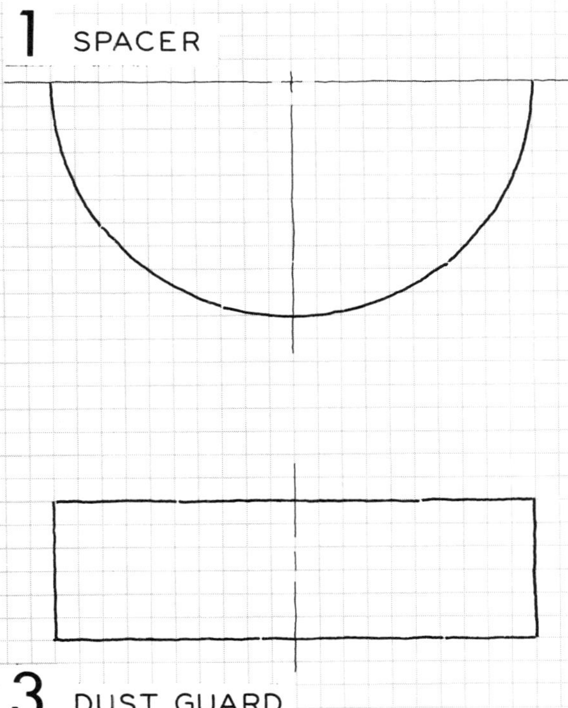

2 PULLEY BLANK

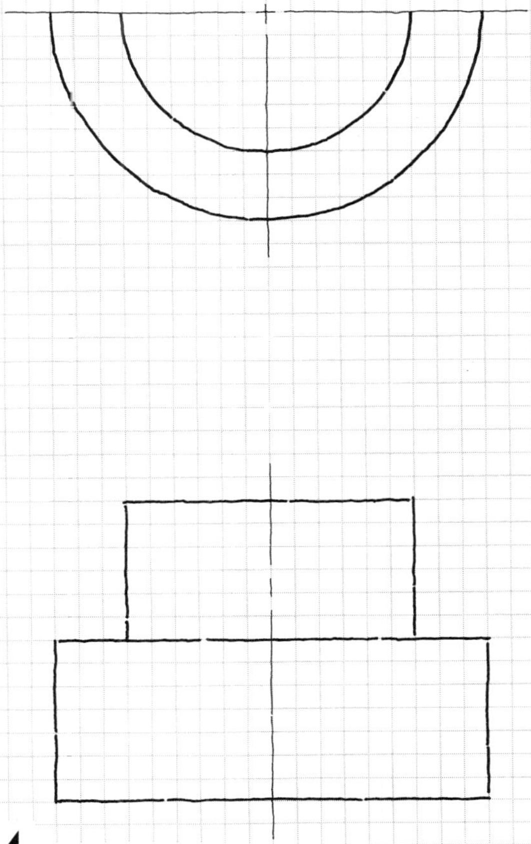

3 DUST GUARD

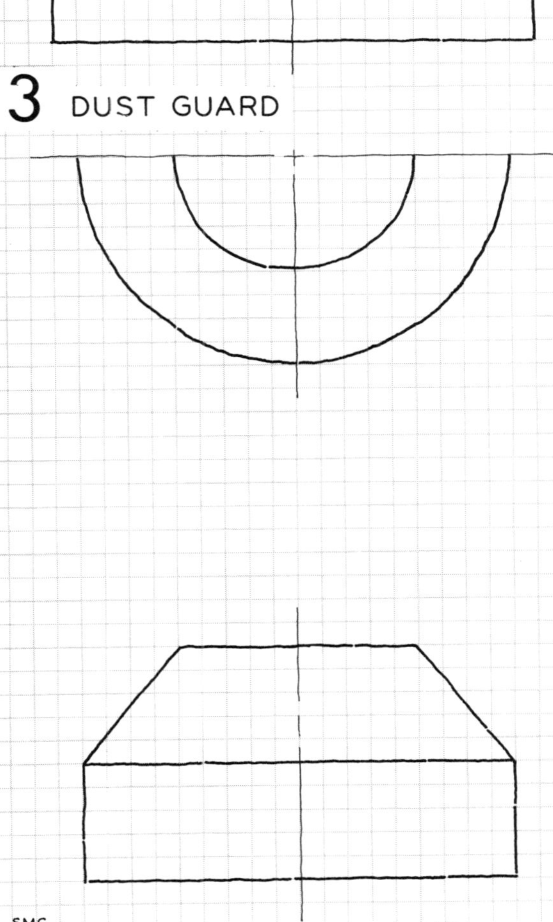

4 SLEEVE

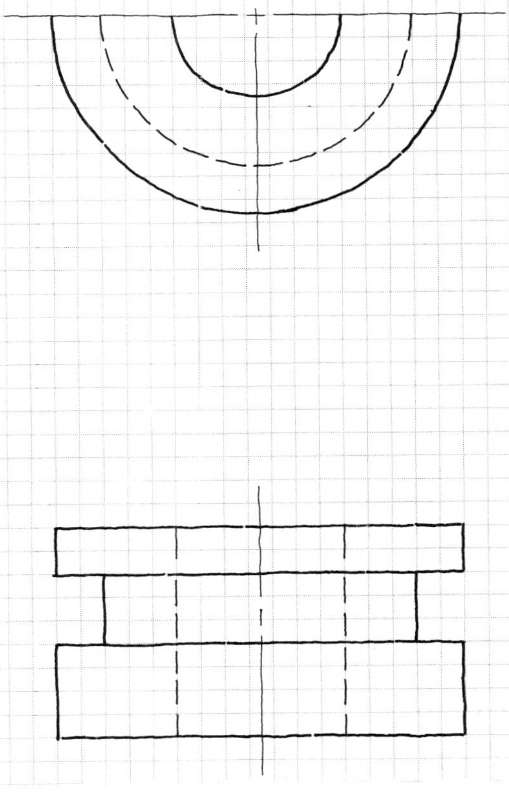

SMC

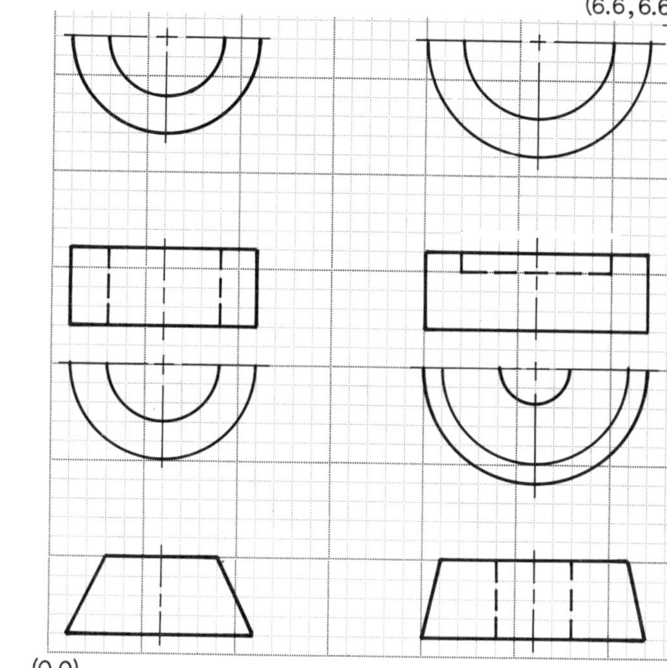


## DIMENSIONING

Plot and dimension the cylindrical and conical parts. Use dimension numerals that are 0.08" high.

# BASIC DIMENSIONING

SKETCH DIMENSIONS FOR THE FULL SIZE OBJECTS.

A. SHOW DIMENSIONS WITH NUMERALS.

B. SHOW DIMENSIONS WITHOUT DIMENSIONS.

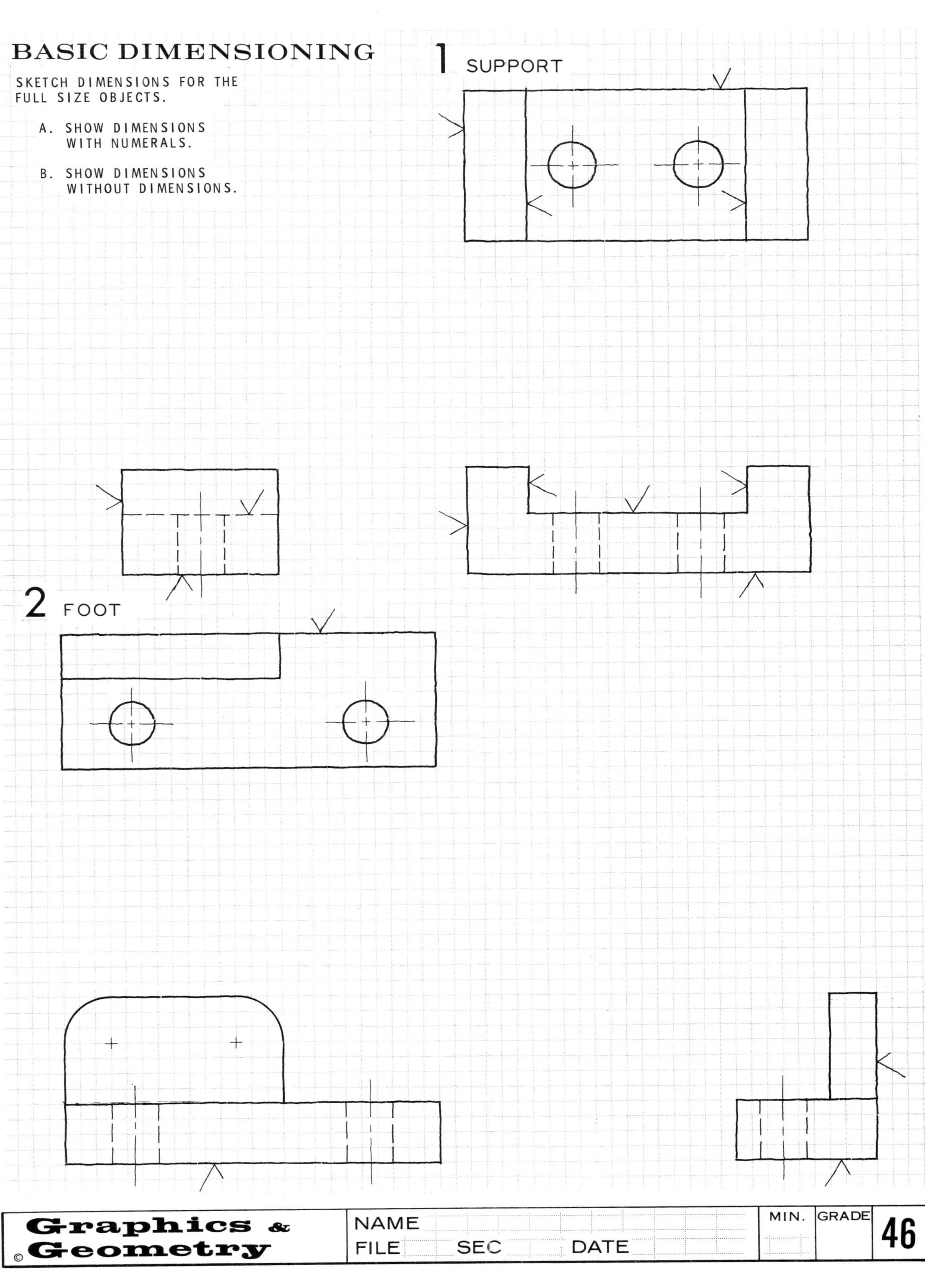

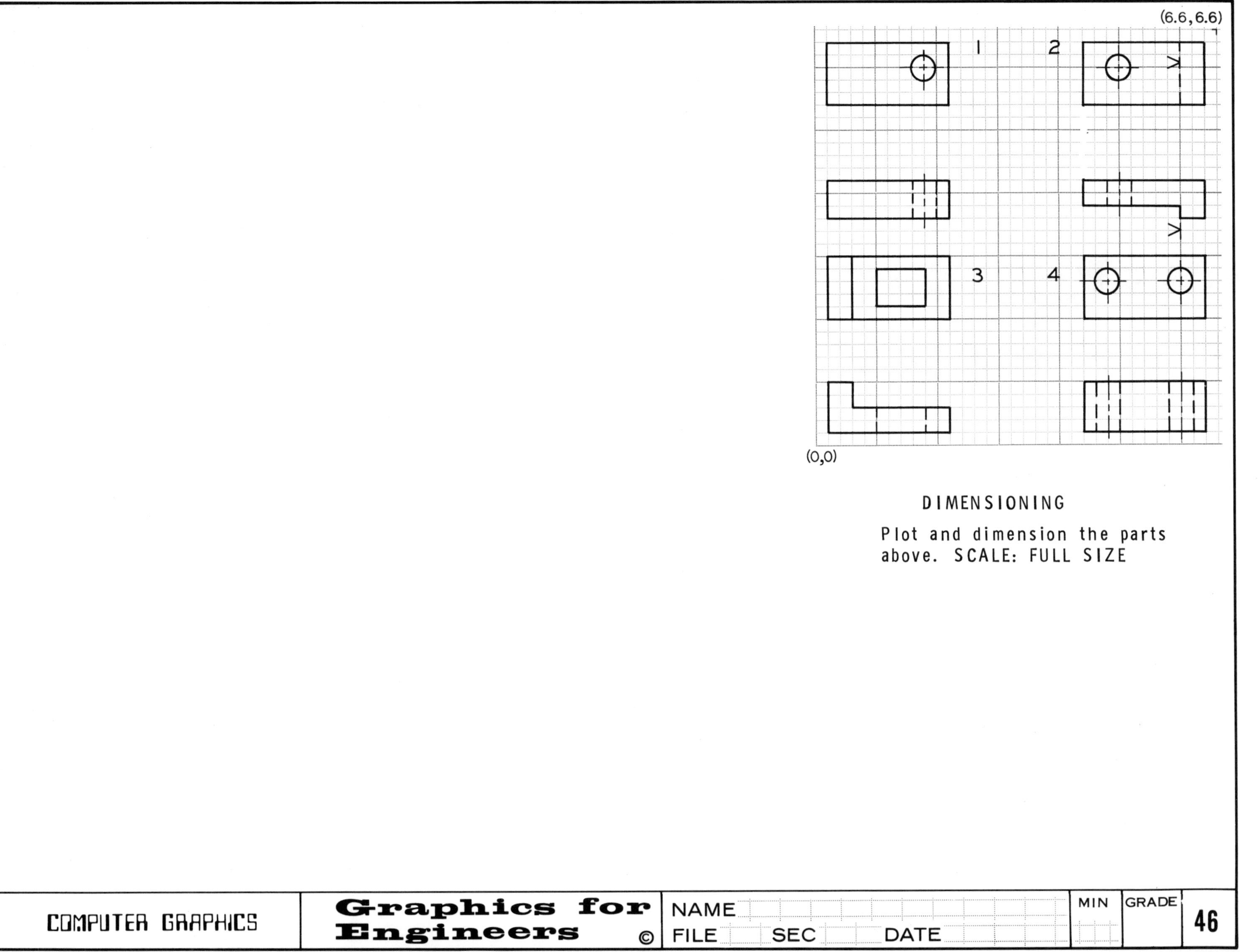


DIMENSIONING

Plot and dimension the parts above. SCALE: FULL SIZE

| COMPUTER GRAPHICS | **Graphics for Engineers** © | NAME<br>FILE SEC DATE | MIN | GRADE | 46 |
|---|---|---|---|---|---|

# DIMENSIONING

DIMENSION THE FULL SIZE PARTS BELOW
USING DECIMAL INCHES OR MILLIMETERS.

1

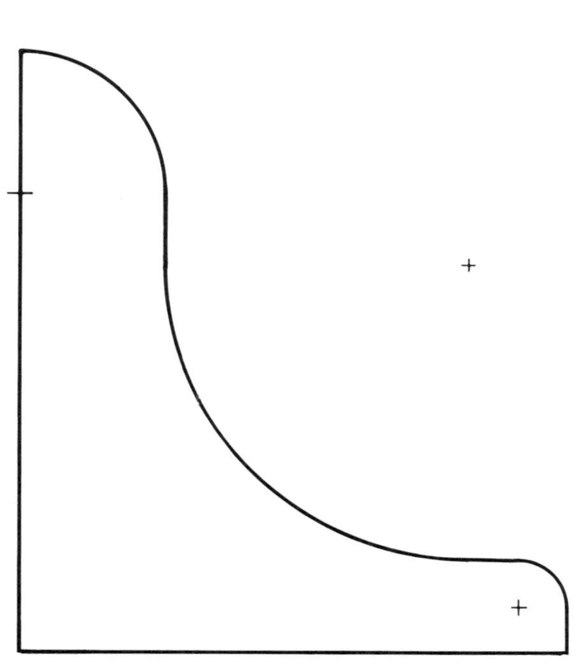

2

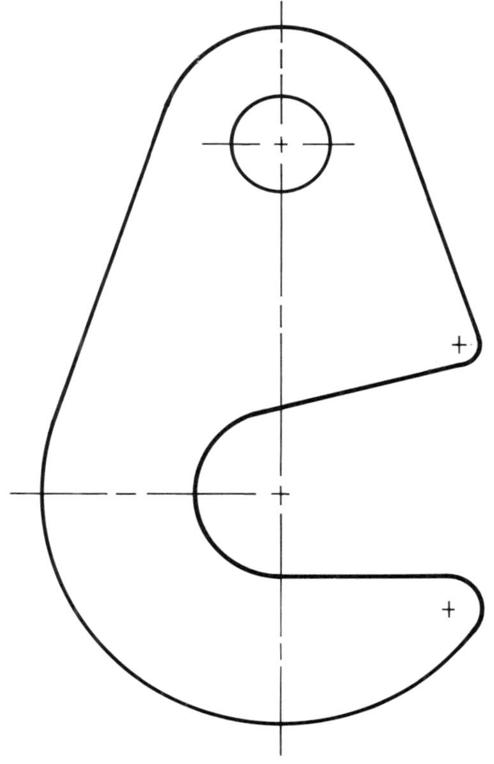

3

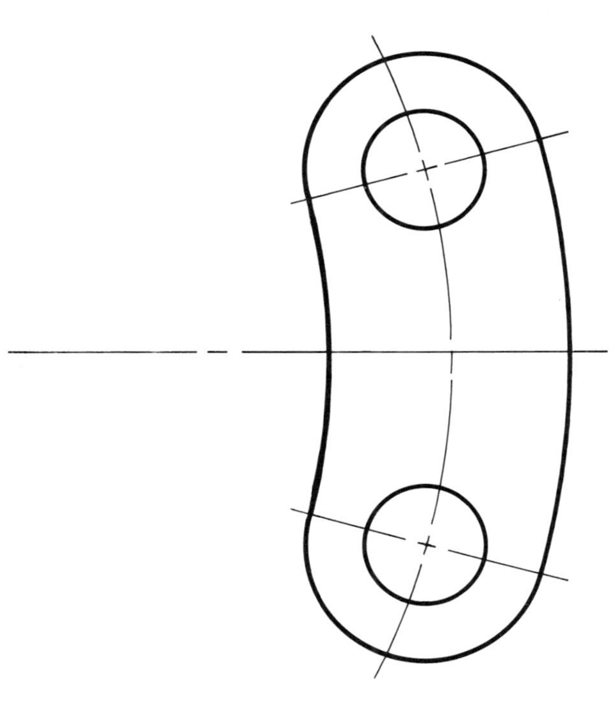

4

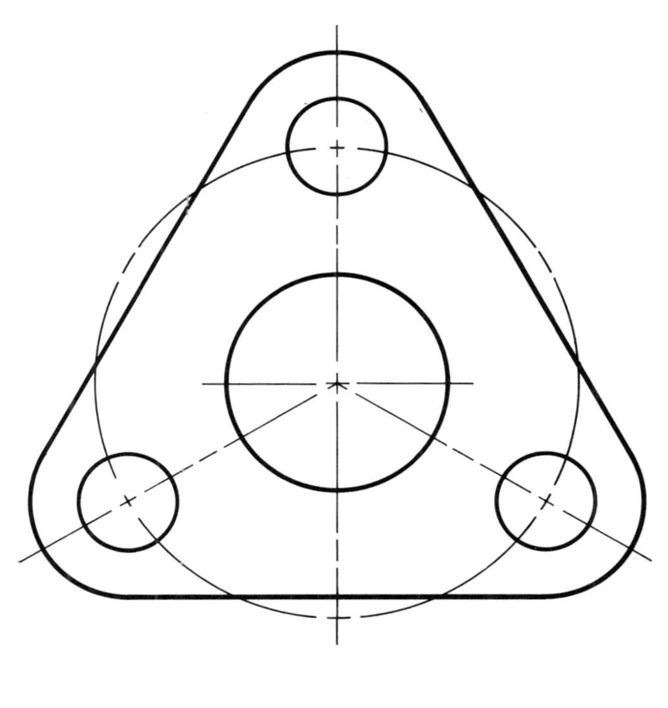

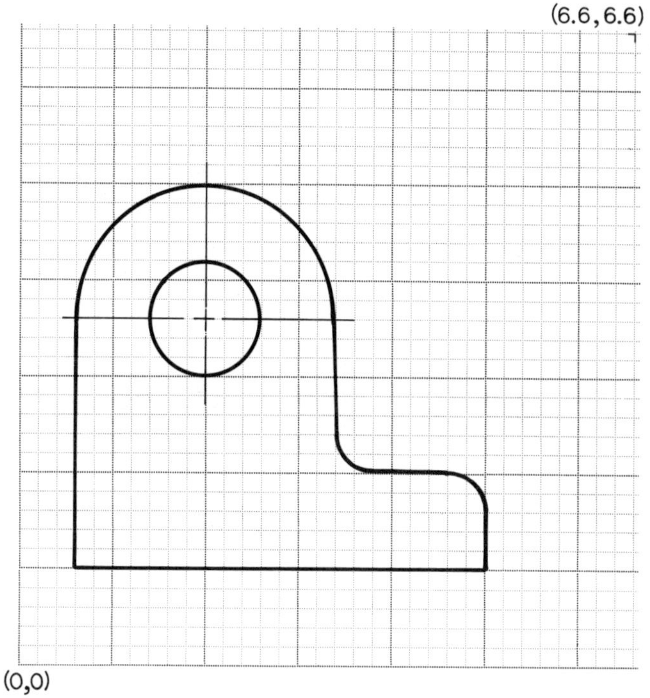

(6.6, 6.6)

(0,0)

DIMENSIONING

Dimension the part given above

SCALE: FULL SIZE

# DIMENSIONS

1

DIMENSION PARTS 1 AND 2 AND THE MACHINED FEATURES OF PROBLEMS 3-8. USE PROBLEM 1 ON THE FOLDOUT SHEETS.

2

3 CHAMFER

4 NECK

5 96 DP KNURL

| Graphics & Geometry | NAME | MIN. | GRADE | 48 |
|---|---|---|---|---|
| | FILE SEC DATE | | | |

©

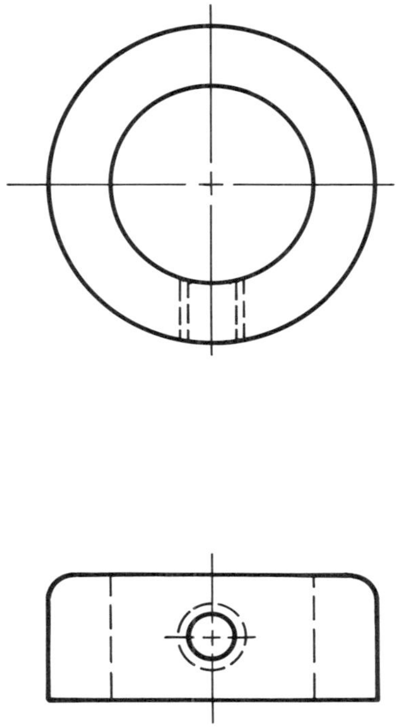

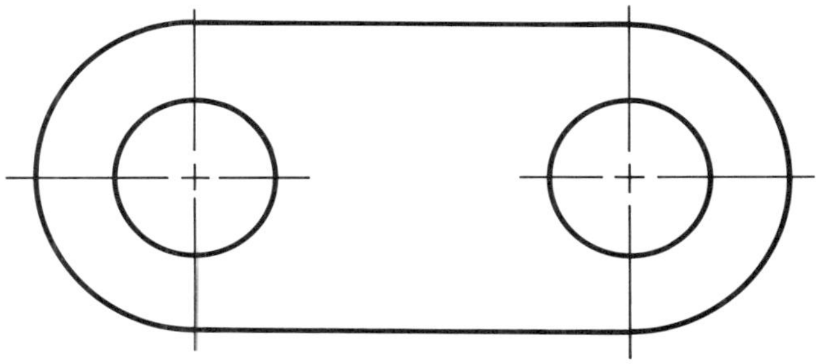

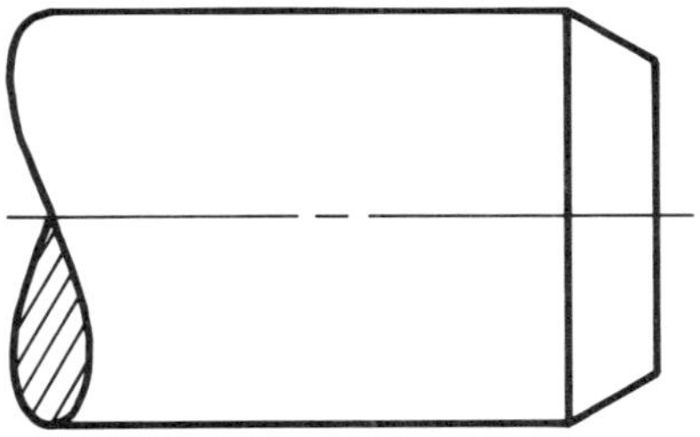

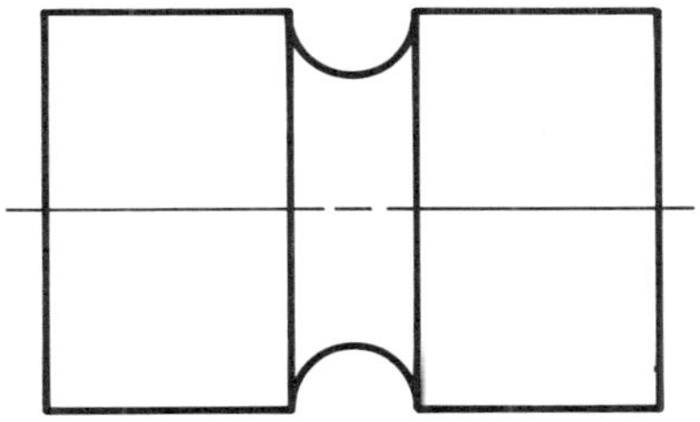

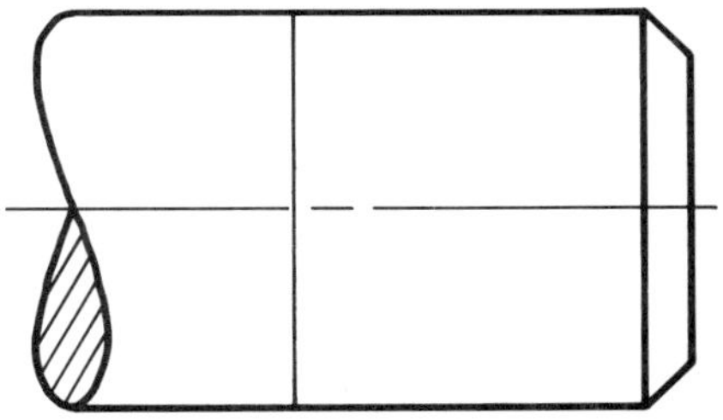

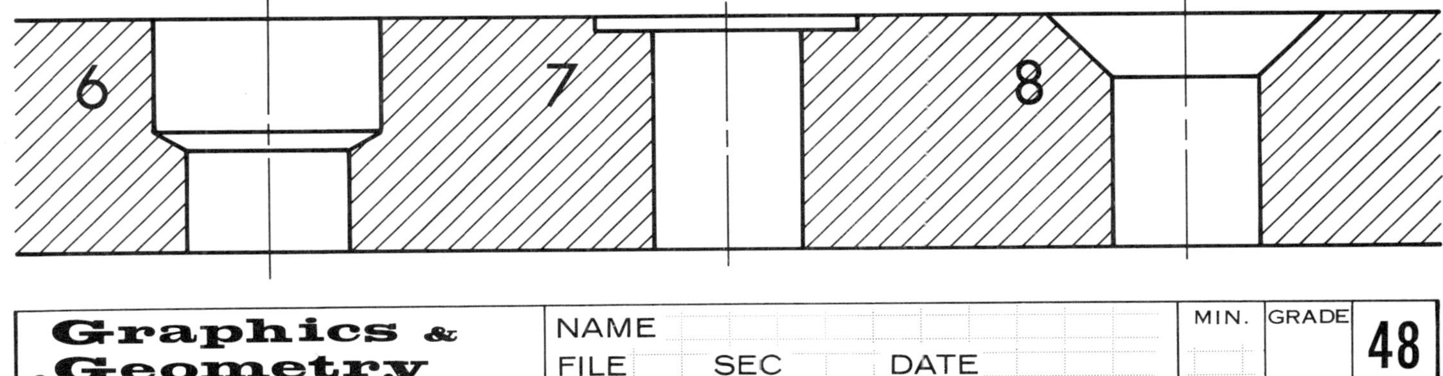

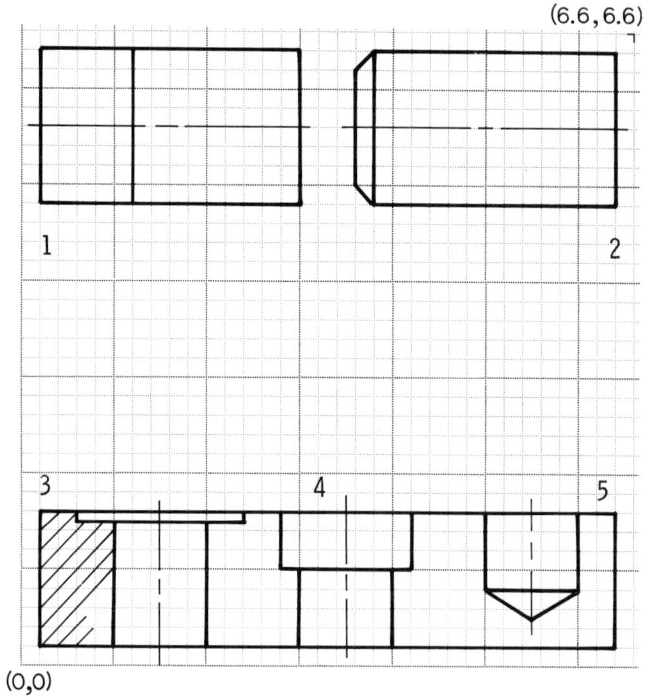

(6.6, 6.6)

1 2

3 4 5

(0,0)

DIMENSIONING NOTES

Dimension the objects above in the following manner:

1. Draw knurl symbols at the right end of the shaft and give a note for a 128 DP DIAMOND KNURL.

2. Dimension the chamfer with a note.

3-5. Dimension the machined holes with notes. Complete the view as a section.

SCALE: FULL SIZE



NAME
FILE SEC DATE
MIN GRADE

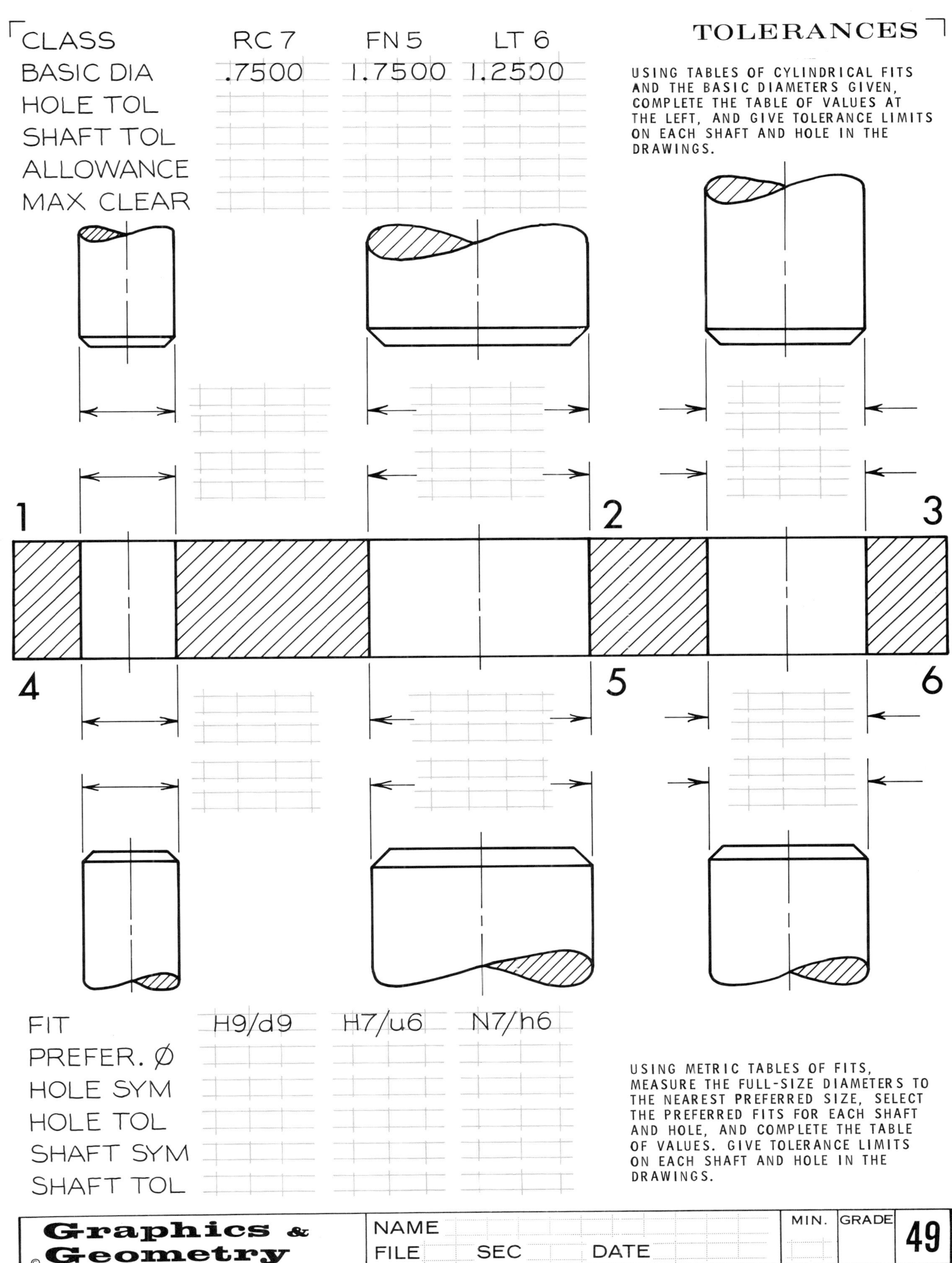
TOLERANCES
USING TABLES OF CYLINDRICAL FITS AND THE BASIC DIAMETERS GIVEN, COMPLETE THE TABLE OF VALUES AT THE LEFT, AND GIVE TOLERANCE LIMITS ON EACH SHAFT AND HOLE IN THE DRAWINGS.
CLASS
RC 7
FN 5
LT 6
BASIC DIA
.7500
1.7500
1.2500
HOLE TOL
SHAFT TOL
ALLOWANCE
MAX CLEAR
1
2
3
4
5
6
FIT
H9/d9
H7/u6
N7/h6
PREFER. Ø
HOLE SYM
HOLE TOL
SHAFT SYM
SHAFT TOL
USING METRIC TABLES OF FITS, MEASURE THE FULL-SIZE DIAMETERS TO THE NEAREST PREFERRED SIZE, SELECT THE PREFERRED FITS FOR EACH SHAFT AND HOLE, AND COMPLETE THE TABLE OF VALUES. GIVE TOLERANCE LIMITS ON EACH SHAFT AND HOLE IN THE DRAWINGS.
Graphics & Geometry
©
NAME
FILE
SEC
DATE
MIN.
GRADE
49

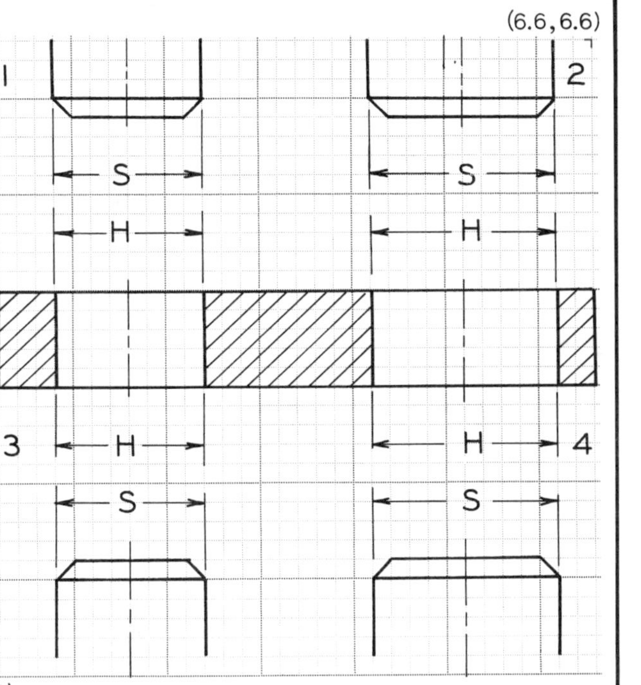


CYLINDRICAL FITS

Plot the views of the shafts (S) and holes (H) as given above. Using ANSI tables for cylindrical fits, compute the limits for the dimensions S and H, and plot them on the drawing in accordance with the following specifications:

1. Basic DIA = 1.60", RC 9 fit, hole basis fit
2. Basic DIA = 2.00", FN 5 fit, hole basis fit
3. Basic DIA = 40 mm, H11/c11 fit, hole basis fit
4. Basic DIA = 50 mm, H7/u6 fit, hole basis fit.

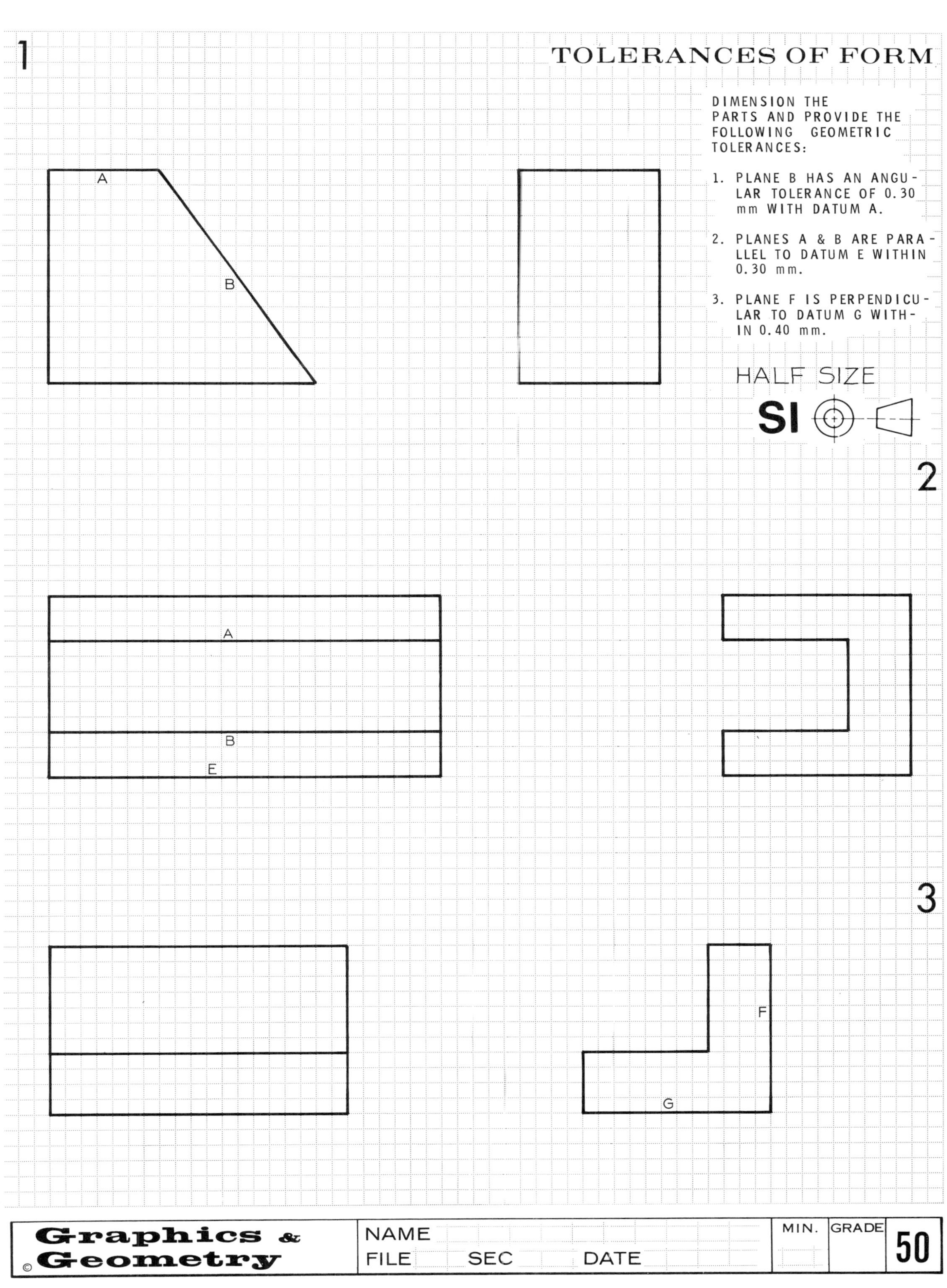
1
TOLERANCES OF FORM
DIMENSION THE PARTS AND PROVIDE THE FOLLOWING GEOMETRIC TOLERANCES:
1. PLANE B HAS AN ANGULAR TOLERANCE OF 0.30 mm WITH DATUM A.
2. PLANES A & B ARE PARALLEL TO DATUM E WITHIN 0.30 mm.
3. PLANE F IS PERPENDICULAR TO DATUM G WITHIN 0.40 mm.
HALF SIZE
SI
A
B
2
A
B
E
3
F
G
Graphics & Geometry
NAME
FILE
SEC
DATE
MIN.
GRADE
50

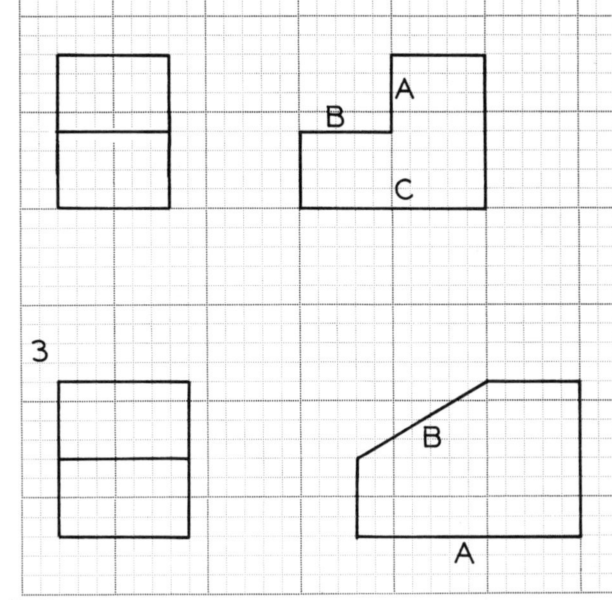


## GEOMETRIC TOLERANCING

Using symbols, note the following features on the parts above:

1. Surface A is perpendicular to datum C within 0.8 mm.
2. Surface B is parallel to datum C within 0.8 mm.
3. Surface B has an angularity tolerance of 0.7 mm with datum A.

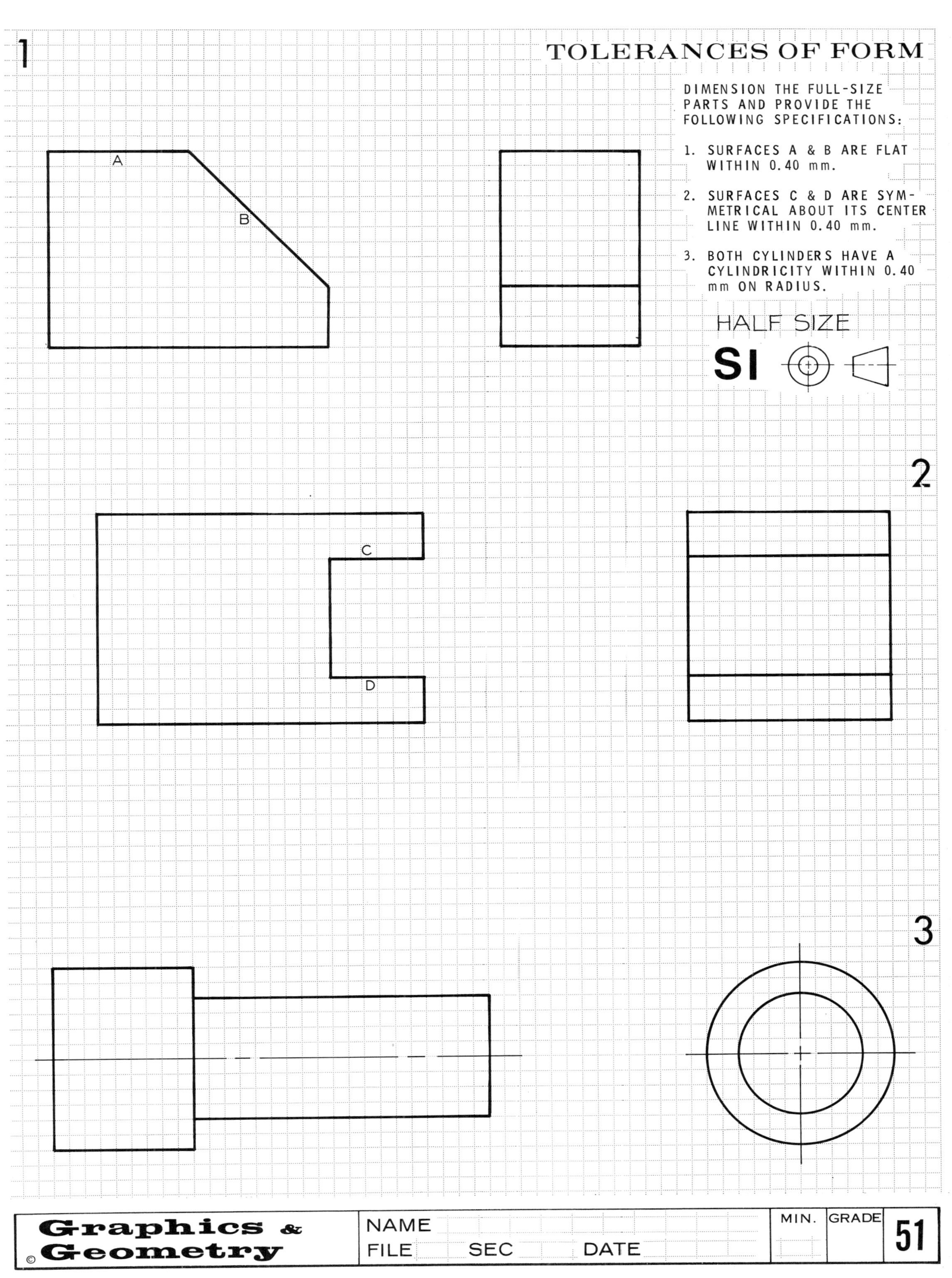
1
TOLERANCES OF FORM
DIMENSION THE FULL-SIZE PARTS AND PROVIDE THE FOLLOWING SPECIFICATIONS:
1. SURFACES A & B ARE FLAT WITHIN 0.40 mm.
2. SURFACES C & D ARE SYMMETRICAL ABOUT ITS CENTER LINE WITHIN 0.40 mm.
3. BOTH CYLINDERS HAVE A CYLINDRICITY WITHIN 0.40 mm ON RADIUS.
HALF SIZE
SI
A
B
2
C
D
3
Graphics & Geometry
NAME
FILE
SEC
DATE
MIN.
GRADE
51

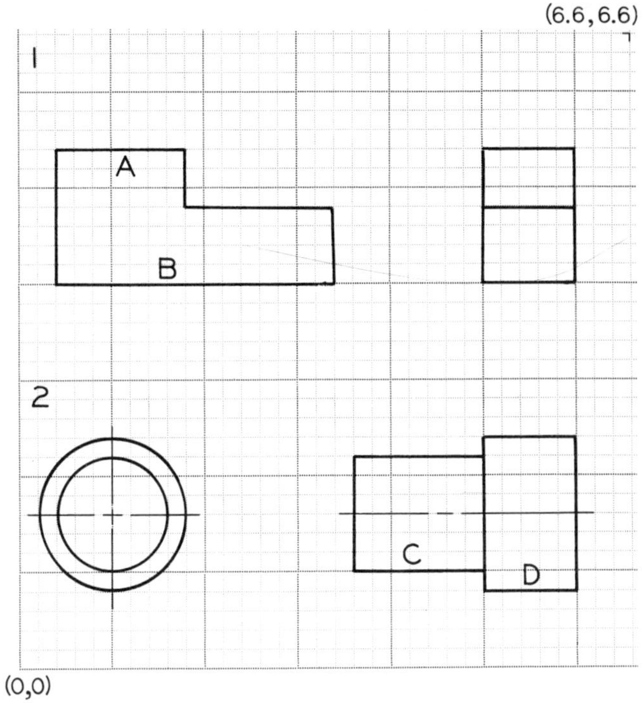
(6.6,6.6)
1
A
B
2
C
D
(0,0)

GEOMETRIC TOLERANCING

Using symbols, note the following features of the parts above:

1. Surfaces A and B are flat within a tolerance of 0.6 mm.
2. Cylinders C and D have a cylindricity tolerance of 0.4 mm.

| | | | | | |
|---|---|---|---|---|---|
| COMPUTER GRAPHICS | **Graphics for Engineers** © | NAME<br>FILE SEC DATE | MIN | GRADE | 51 |

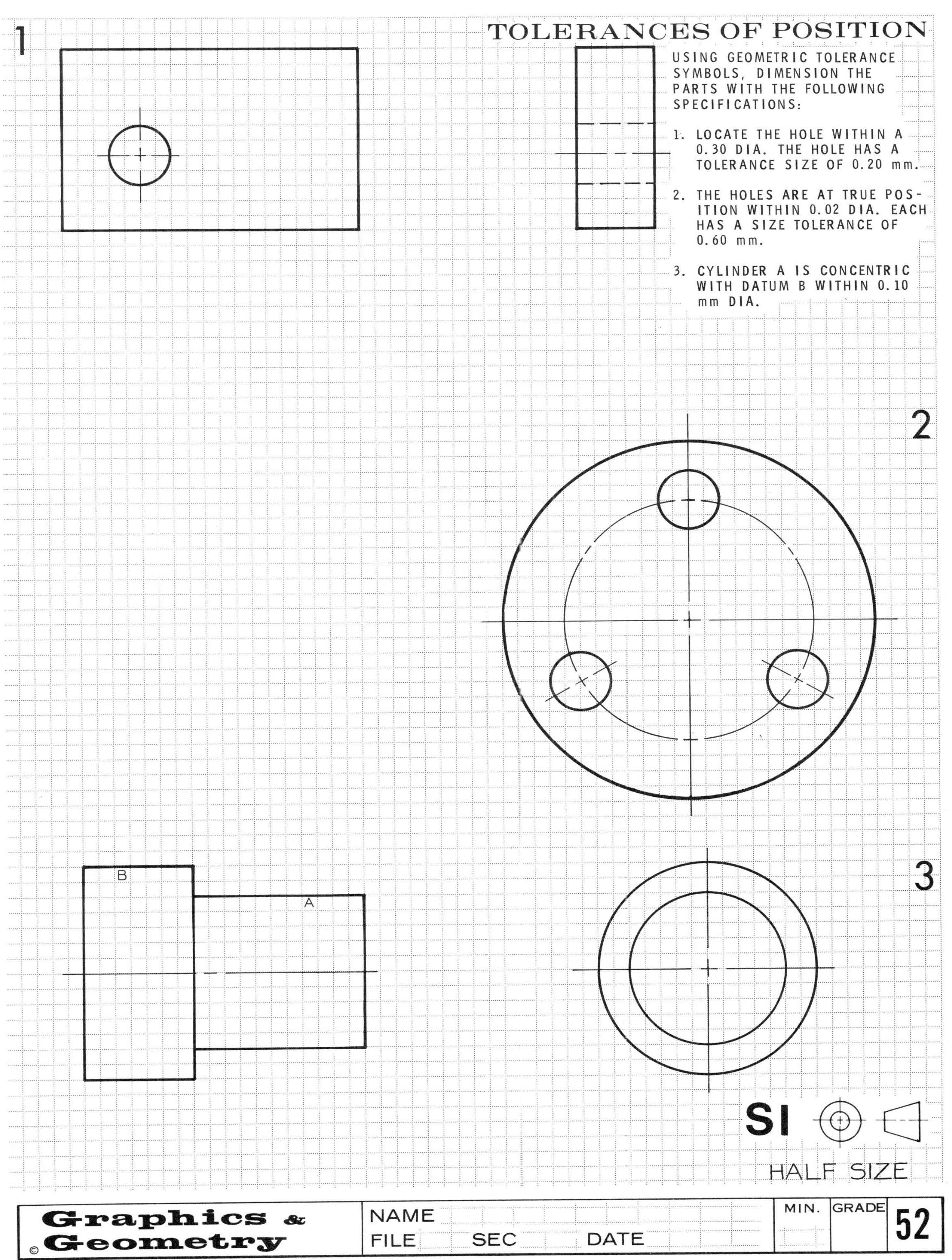
1
TOLERANCES OF POSITION
USING GEOMETRIC TOLERANCE SYMBOLS, DIMENSION THE PARTS WITH THE FOLLOWING SPECIFICATIONS:
1. LOCATE THE HOLE WITHIN A 0.30 DIA. THE HOLE HAS A TOLERANCE SIZE OF 0.20 mm.
2. THE HOLES ARE AT TRUE POSITION WITHIN 0.02 DIA. EACH HAS A SIZE TOLERANCE OF 0.60 mm.
3. CYLINDER A IS CONCENTRIC WITH DATUM B WITHIN 0.10 mm DIA.
2
B
A
3
SI
HALF SIZE
Graphics & Geometry
NAME
FILE
SEC
DATE
MIN.
GRADE
52

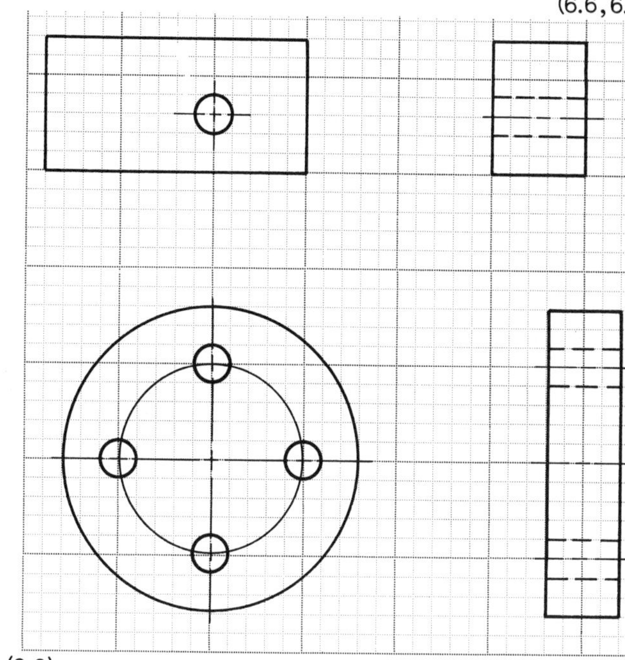


GEOMETRIC TOLERANCING

1. Locate the hole at true position within a 0.4 mm DIA. The hole tolerance is 0.6 mm.

2. Locate the four holes at true position within a 0.6 mm DIA. The holes have a size tolerance of 0.4 mm .

# TOLERANCES

1 USING THE RECOMMENDED TOLERANCES FOR A MEDIUM SERIES, GIVE THE UPPER AND LOWER LIMITS BASED ON THE 52 mm BASIC DIMENSION.

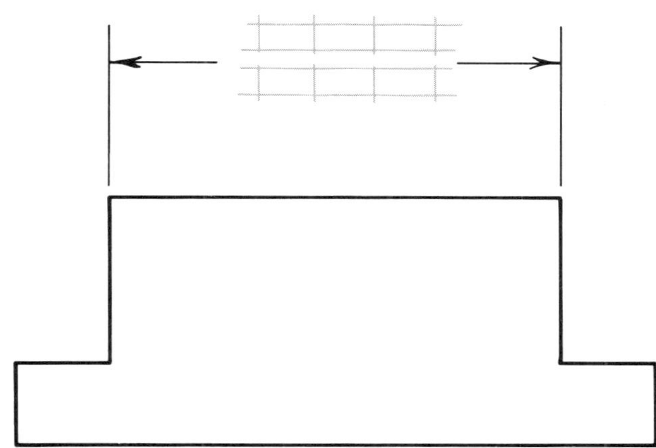

USING THE IT (INTERNATIONAL TOLERANCES) TABLE, GIVE THE UPPER AND LOWER LIMITS BASED ON A 52 mm BASIC DIMENSION. USE THE MAXIMUM IT GRADE OBTAINED BY THE SURFACE GRINDING PROCESS. 2

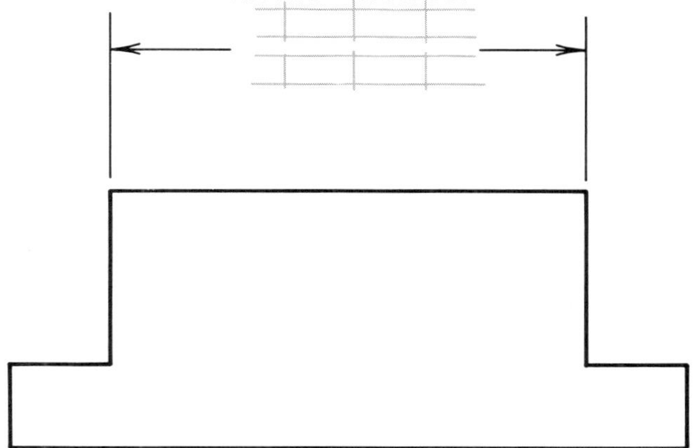

3 USING THE RECOMMENDED TOLERANCES FOR ANGLES IN DEGREES AND BY MEASURING THE SIDE OF THE ANGLE (DRAWN FULL SIZE), GIVE THE TOLERANCE OF THE ANGLE IN A PLUS-AND-MINUS FORM.

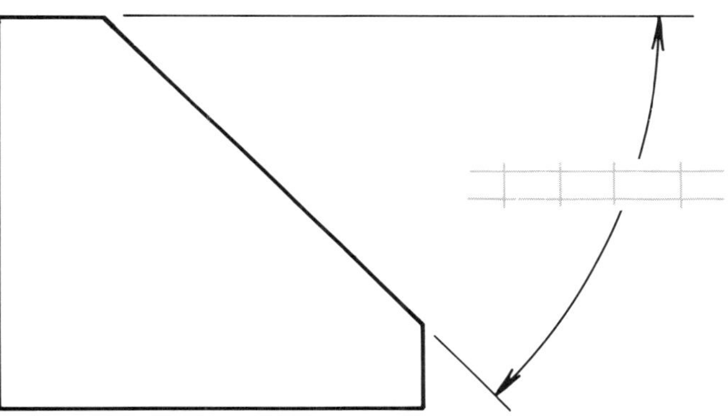

BY REFERRING TO THE RECOMMENDED GENERAL TOLERANCES TABLES, COMPLETE THE GENERAL TOLERANCE TABLE FOR A FINE SERIES FOR THE SIZES GIVEN. 4

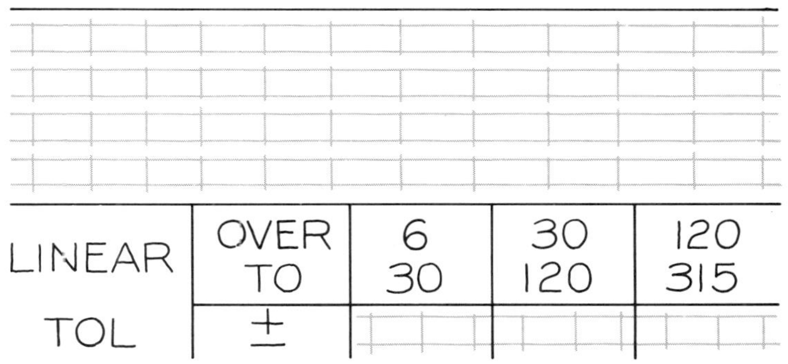

DIMENSIONS IN mm

| LINEAR | OVER<br>TO | 6<br>30 | 30<br>120 | 120<br>315 |
|---|---|---|---|---|
| TOL | ± | | | |

5 COMPLETE THE SPECIFICATIONS OF THE SURFACE TEXTURE NOTES BELOW. SURFACE A IS TO BE FINISHED BY REMOVING 2.0 mm OF MATERIAL. SURFACE B IS TO BE FINISHED BY GRINDING.

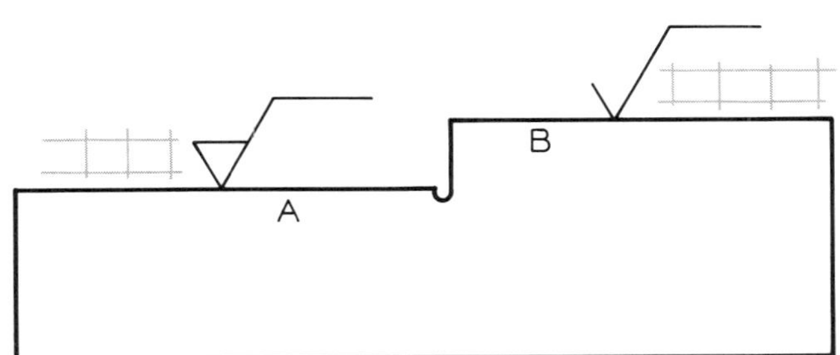


SURFACE C HAS A ROUGHNESS AVERAGE RATING OF 2.0 MICROMETERS, A LAY PERPENDICULAR TO THE SURFACE, AND A ROUGHNESS SPACING OF 1.0 mm. SURFACE D HAS THE ROUGHNESS AVERAGE LIMITS RESULTING FROM SURFACE GRINDING, A MAXIMUM WAVINESS HEIGHT OF 0.040 AND A SPACING OF 8 mm. LAY IS PARALLEL TO THE SURFACE. 6

C

D

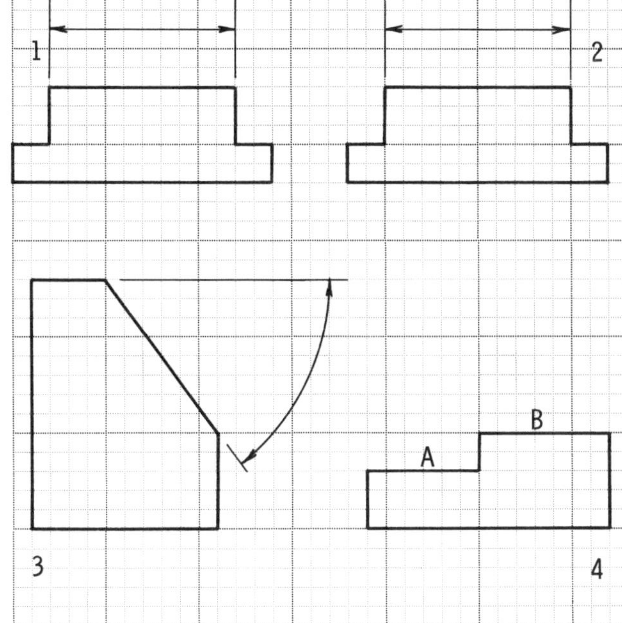


TOLERANCES

1. Using the recommended tolerances for a medium series, give the upper and lower limits based on a 52 mm basic dimension.
2. Using the IT table, give the upper and lower limits based on a 52 mm
q basic dimension. Use the max IT grade obtained by surface grinding.
3. Using the recommended tolerances forangles and by measuring the side of the angle (drawn full size), give the tolerance of the angle in plus-and-minus form.
4. Give a surface texture note on A that is to be finished by removing 2 mm of material. B is to be finished by grinding.

# DIMENSIONING

SHOWN ARE THREE VIEWS OF A TRUNNION, A JIG AND FIXTURE PART. DIMENSION THE VIEWS COMPLETELY AS ASSIGNED OR BY ONE OF THE FOLLOWING METHODS. USE THE 1/8" GRIDS AS A FULL SIZE SCALE.

A. SKETCH THE DIMENSIONS ON THIS SHEET, OBSERVING PROPER SPACING OF DIMENSION LINES.

B. USING INSTRUMENTS, DIMENSION THE VIEWS.

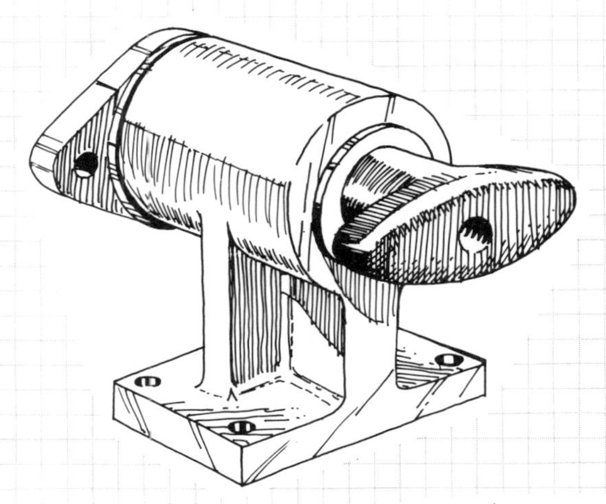

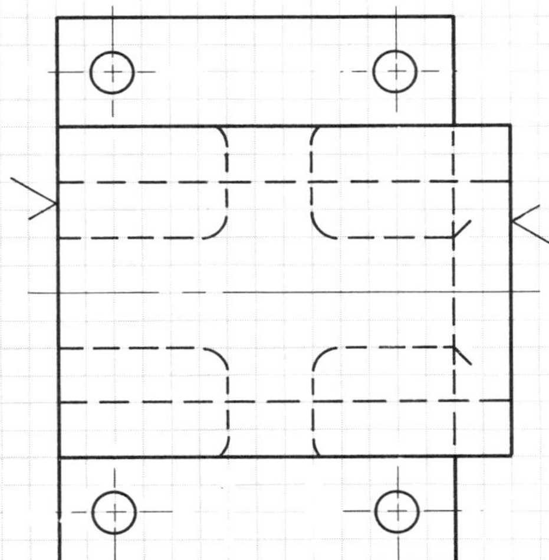

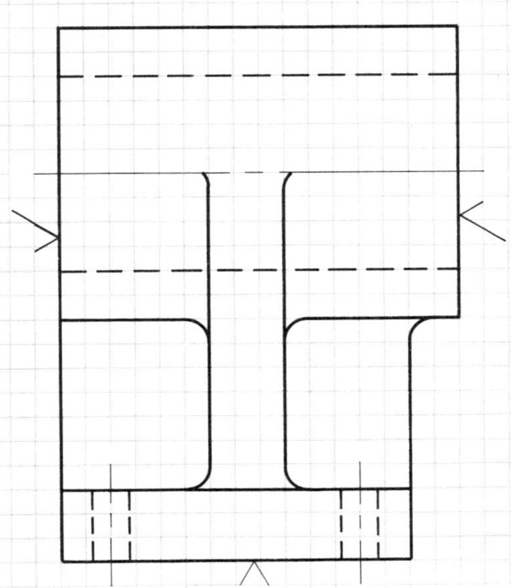

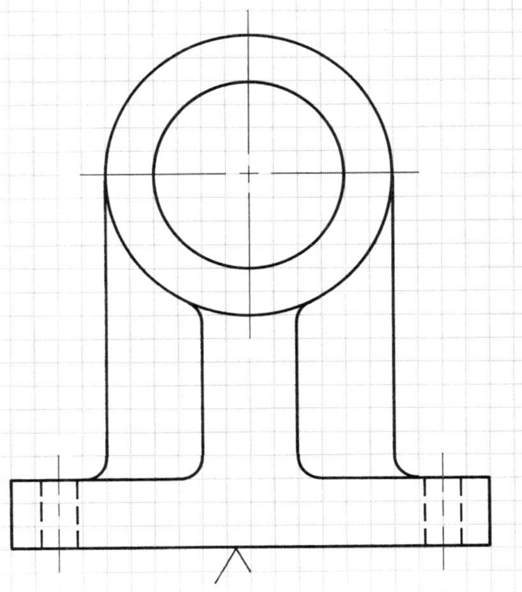

## DIMENSIONING

Plot and dimension the Jig Fixture in accordance with the specifications on the opposite side of this sheet.

DIMENSION THE CENTER TRUNION USING SI UNITS. APPLY THE FOLLOWING TOLERANCES: THE TWO 16 mm DIA PINS--+0.00 -0.05; THE 51 mm DIA HOLE--+0.05 -0.00 mm; THE FOUR THREADED HOLES ARE M 12 X 1.75. MEASURE AND DIMENSION ALL OTHER DIMENSIONS TO THE NEAREST MILLIMETER.

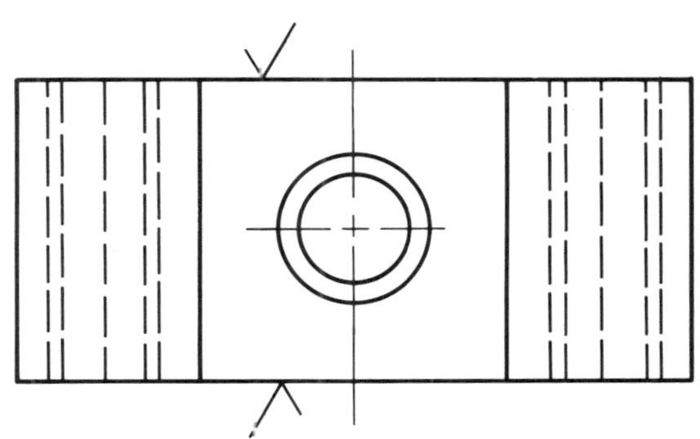

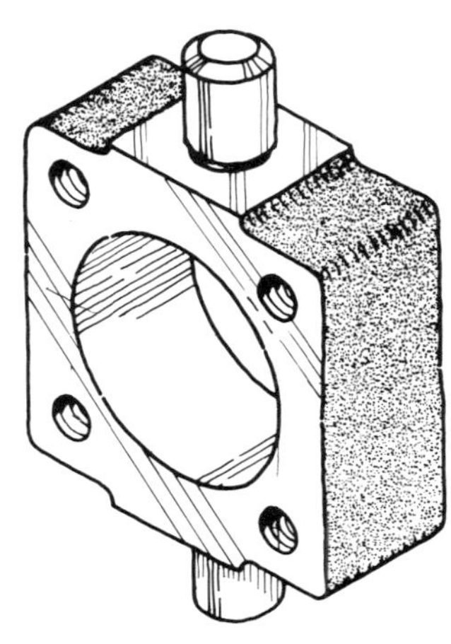

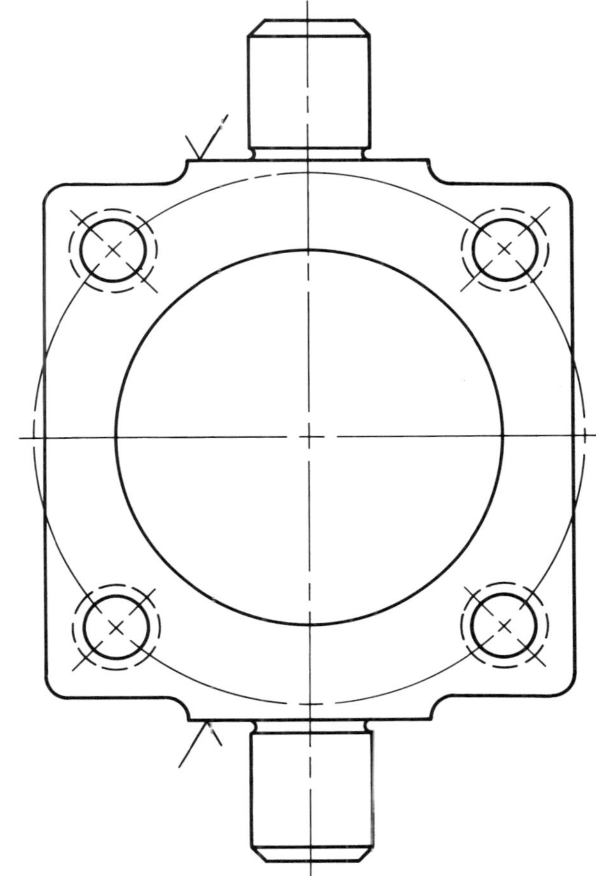

CENTER TRUNION

SCALE: 1:1

FILLETS & ROUNDS 4R

TOLERANCES

UNSPECIFIED DIMENSIONS ±0.5

ANGLES ±0.5°

## WORKING DRAWINGS

Plot and dimension the Center Trunion in accordance with the specifications given on the opposite side of this sheet.

## Working Drawings - See fold-out sheets at rear

CONVERT THE FREEHAND SKETCHES BELOW INTO WORKING DRAWINGS MADE WITH INSTRUMENTS ON THE FOLDOUT VELLUM SHEETS AT THE REAR OF THE BOOK. DIMENSIONS GIVEN ON THE SKETCHES MAY BE INCORRECTLY POSITIONED. CORRECT ALL ERRORS THAT YOU MIGHT FIND.

PARTS 1, 2, 3, & 4 HAVE BEEN DRAWN TO SCALE ON EARLIER SHEETS. BY SLIGHTLY REARRANGING THESE VIEWS, YOU WILL BE ABLE TO TRACE THEM ONTO THE VELLUM SHEETS AT THE REAR OF THIS BOOK.

CONSTRUCT AN ASSEMBLY OF THE DRIVE TENSIONER ON THE THIRD FOLD-OUT SHEET AND GIVE A PARTS LIST.

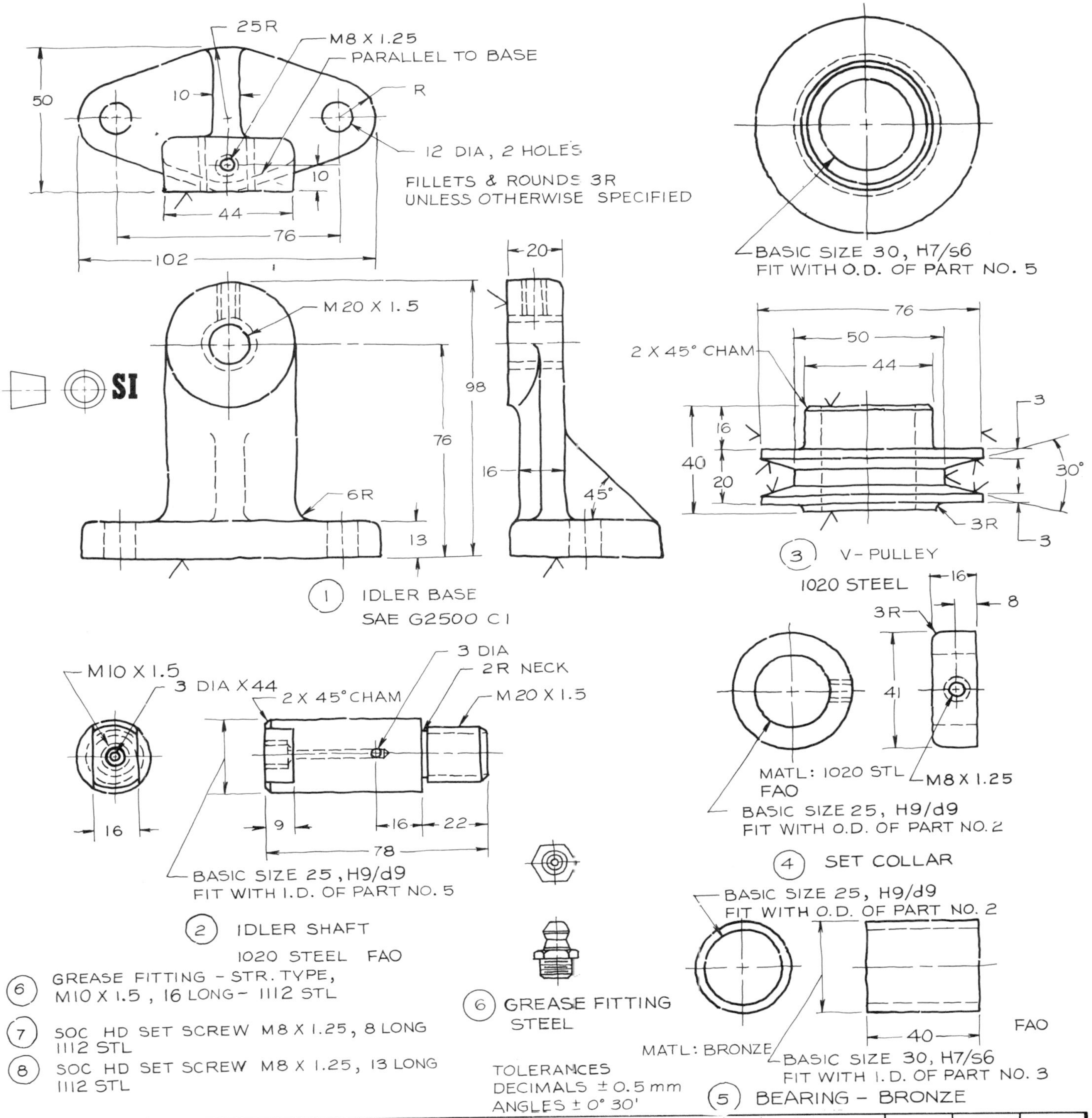

25R
M8 X 1.25
PARALLEL TO BASE
50
10
R
12 DIA, 2 HOLES
10
FILLETS & ROUNDS 3R
UNLESS OTHERWISE SPECIFIED
44
76
102
20
M 20 X 1.5
SI
98
76
16
45°
6R
13

(1) IDLER BASE
SAE G2500 C I

BASIC SIZE 30, H7/s6
FIT WITH O.D. OF PART NO. 5
76
50
2 X 45° CHAM
44
3
16
40
20
30°
3R
3

(3) V-PULLEY
1020 STEEL

M10 X 1.5
3 DIA X 44
3 DIA
2R NECK
2 X 45° CHAM
M 20 X 1.5
16
9
16
22
78
BASIC SIZE 25, H9/d9
FIT WITH I.D. OF PART NO. 5

(2) IDLER SHAFT
1020 STEEL FAO

16
8
3R
41
MATL: 1020 STL
FAO
M8 X 1.25
BASIC SIZE 25, H9/d9
FIT WITH O.D. OF PART NO. 2

(4) SET COLLAR

BASIC SIZE 25, H9/d9
FIT WITH O.D. OF PART NO. 2
40
FAO
MATL: BRONZE
BASIC SIZE 30, H7/S6
FIT WITH I.D. OF PART NO. 3

(5) BEARING - BRONZE

(6) GREASE FITTING
STEEL

(6) GREASE FITTING - STR. TYPE, M10 X 1.5, 16 LONG - 1112 STL

(7) SOC HD SET SCREW M8 X 1.25, 8 LONG 1112 STL

(8) SOC HD SET SCREW M8 X 1.25, 13 LONG 1112 STL

TOLERANCES
DECIMALS ± 0.5 mm
ANGLES ± 0° 30'

| Graphics & Geometry | NAME | MIN. | GRADE | 56 |
|---|---|---|---|---|
| © | FILE SEC DATE | | | |

## OBLIQUE DRAWING

1 PROBLEM 1: DRAW DOUBLE SIZE OBLIQUES OF THE BOX PARALLELS IN THE SPACES PROVIDED.

PROBLEM 2: DRAW A DOUBLE SIZE CABINET OBLIQUE OF THE ANGLE BRACKET.

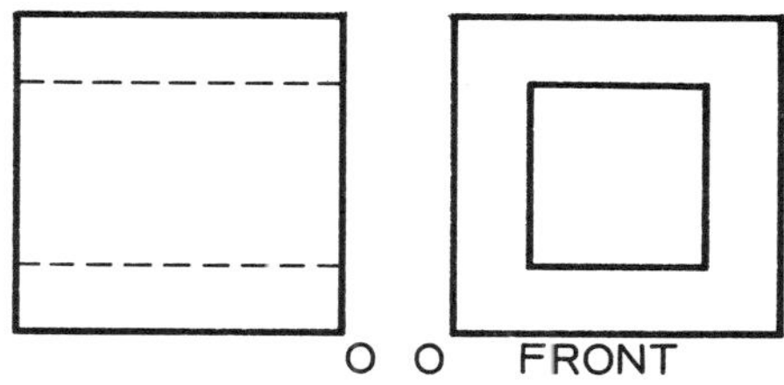

COURTESY TAFT - PEIRCE CO.

CAVALIER - FRONT

CABINET - FRONT

2

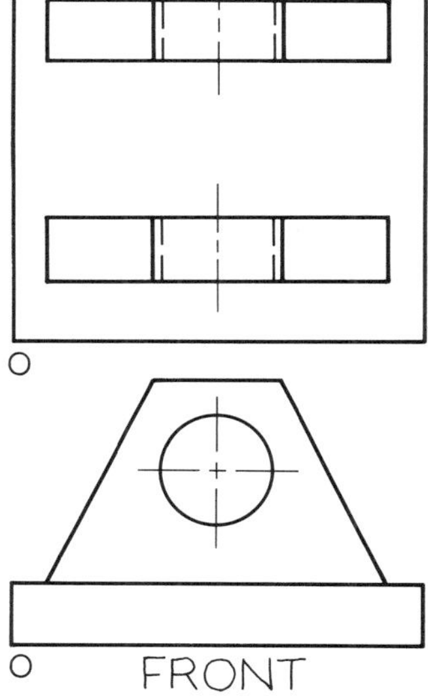

COURTESY OF HYDRO-LINE CO.

FRONT

| Graphics & Geometry | NAME | MIN. | GRADE | 57 |
|---|---|---|---|---|
| | FILE SEC DATE | | | |

©

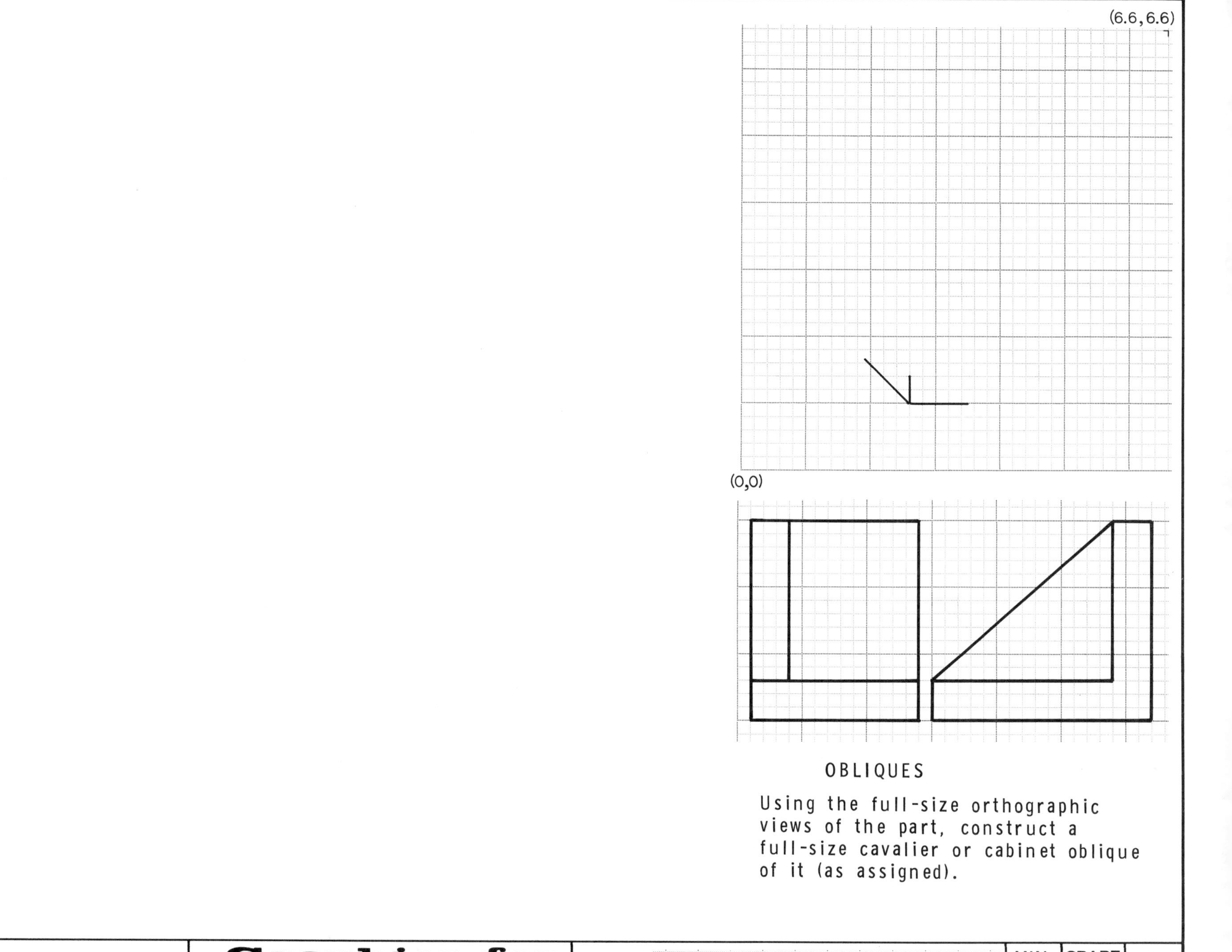


## OBLIQUES

Using the full-size orthographic views of the part, construct a full-size cavalier or cabinet oblique of it (as assigned).

DOUBLE THE DIMENSIONS OF THE ORTHOGRAPHIC VIEWS AND DRAW OBLIQUES OF THE OBJECTS IN THE SPACES INDICATED.

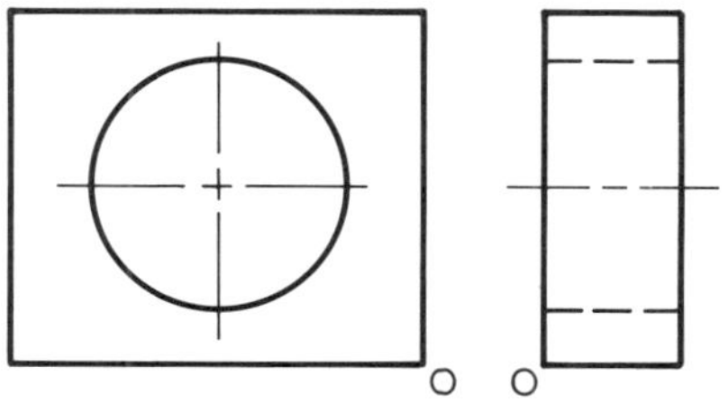

# OBLIQUES

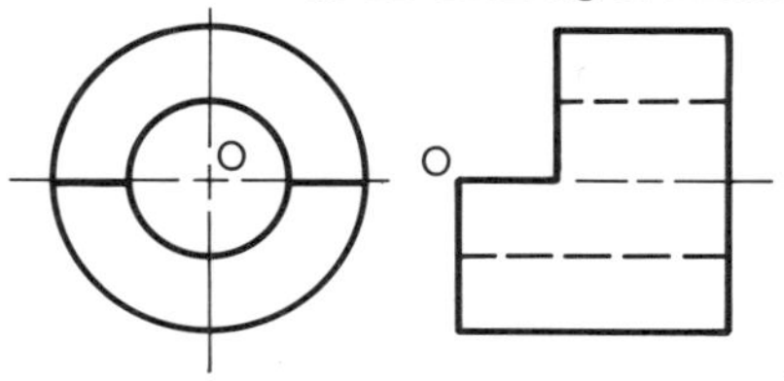

1

2

CAVALIER

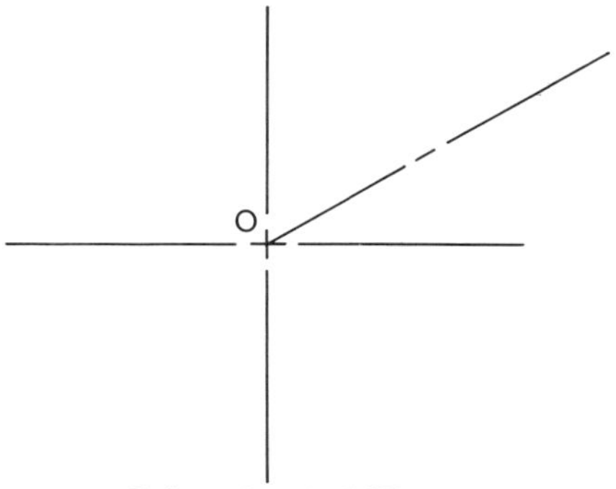

CAVALIER

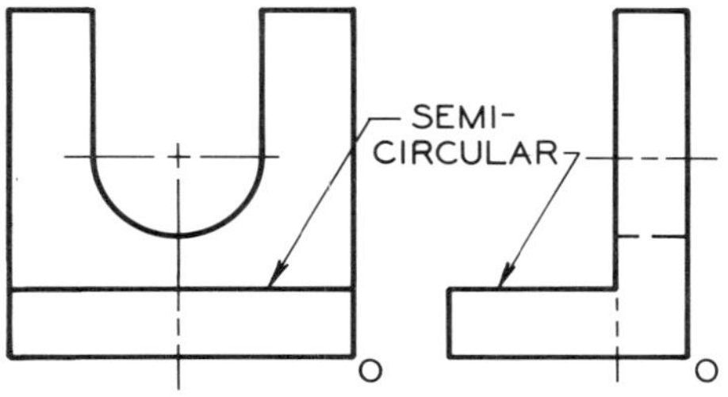

SEMI-CIRCULAR

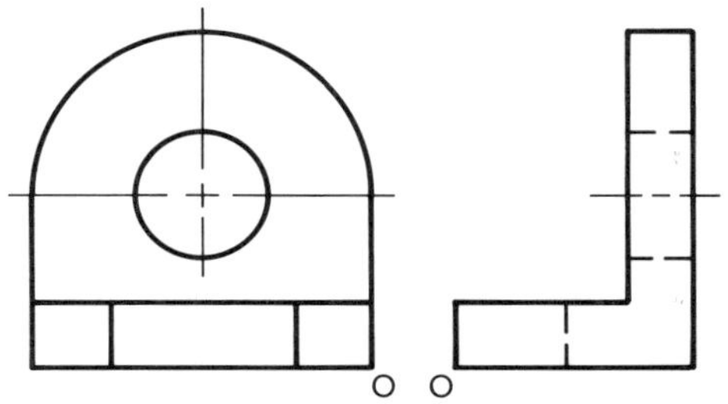

3

4

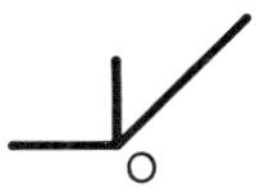

CAVALIER

CABINET

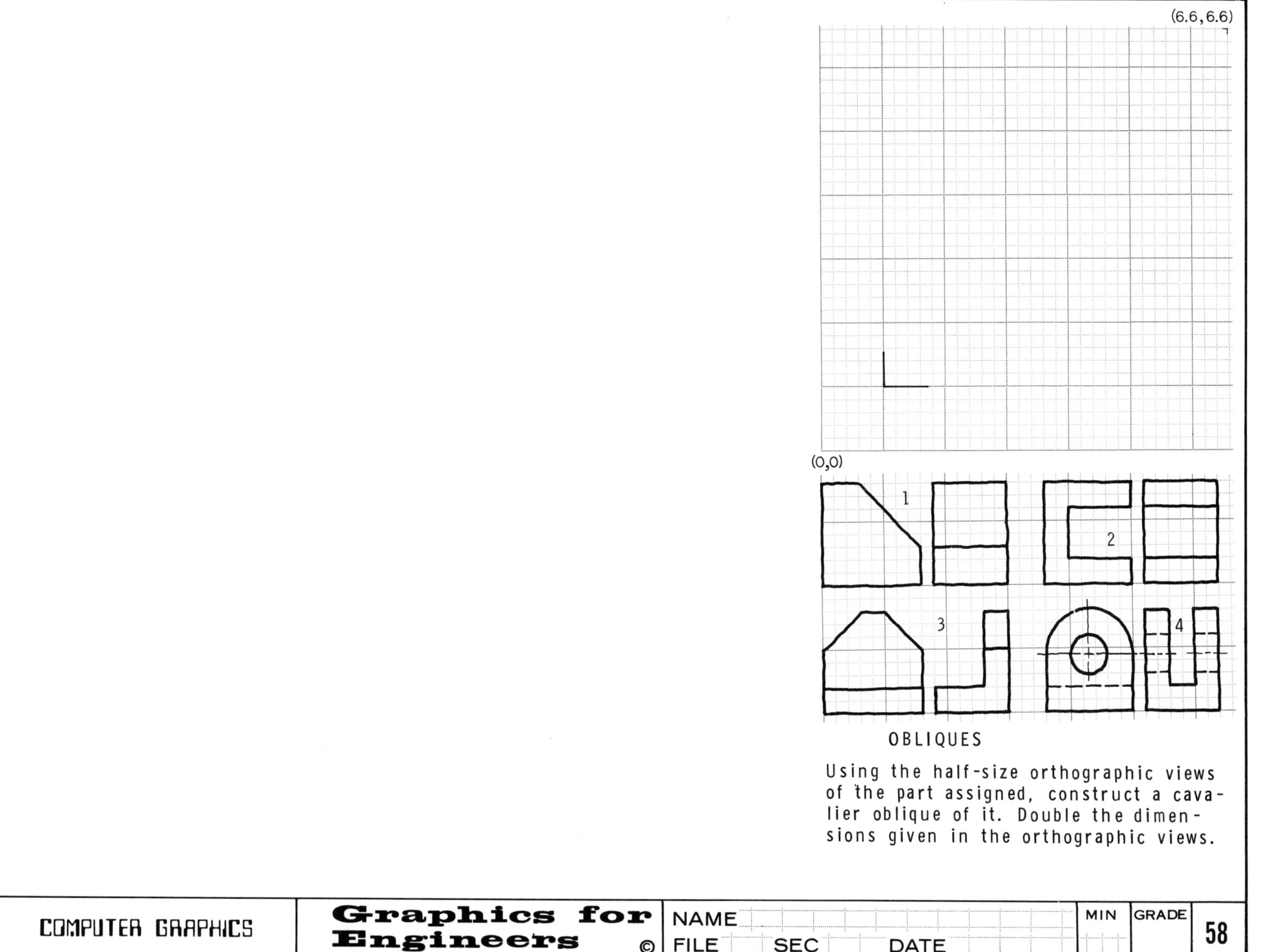


## OBLIQUES

Using the half-size orthographic views of the part assigned, construct a cavalier oblique of it. Double the dimensions given in the orthographic views.

# OBLIQUES

1 USING INSTRUMENTS, DRAW DOUBLE SIZE CAVALIER AND CABINET OBLIQUES OF THE CAST T SECTION.

PROB. 2: DRAW A DOUBLE SIZE CABINET OBLIQUE OF THE ANGLE BRACKET.

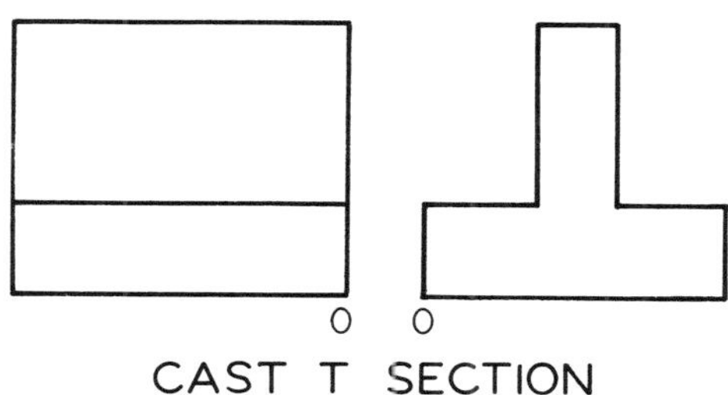

CAST T SECTION

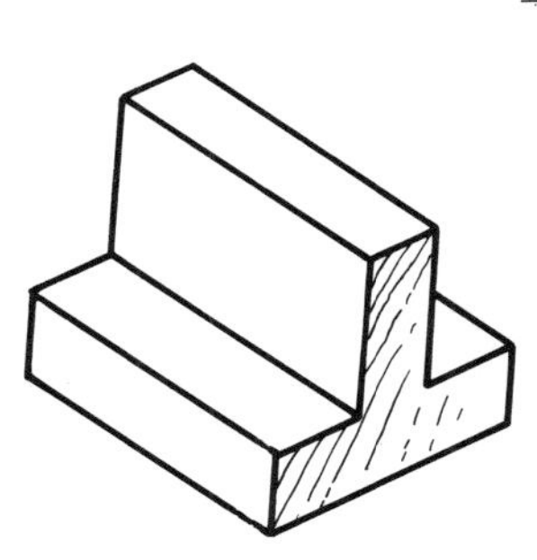

CAVALIER

CABINET

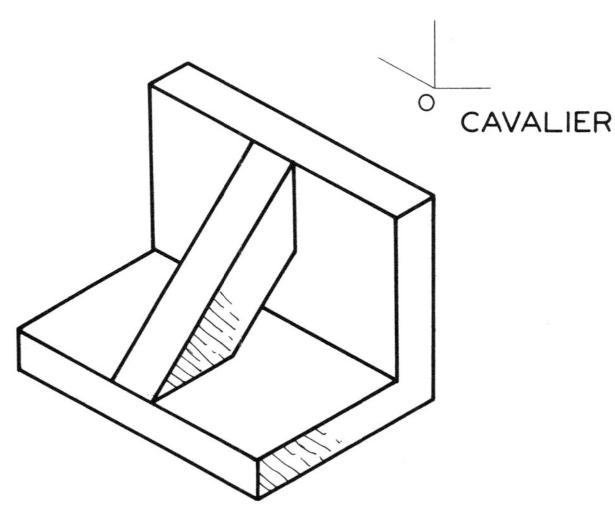

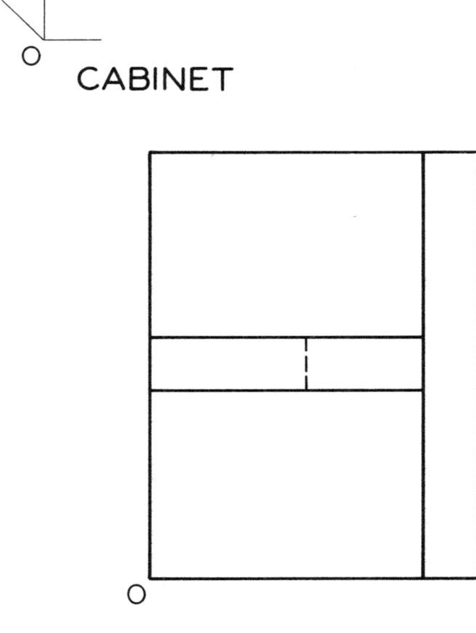

2

GUSSETED ANGLE BRACKET



©

NAME

FILE SEC DATE

MIN. GRADE

(0,0)

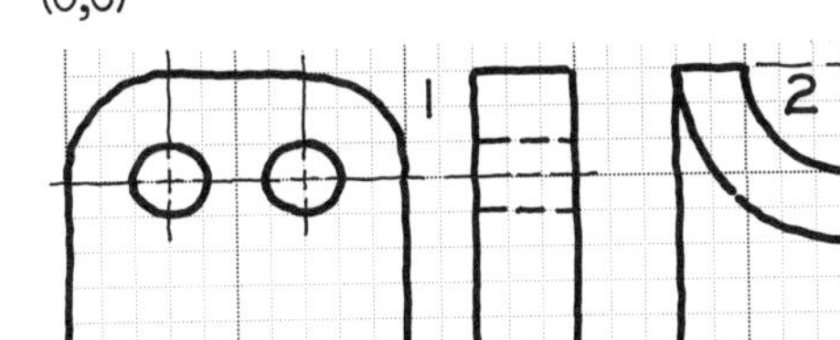


ISOMETRICS

Using the half-size orthographic views of the part assigned, construct a full-size isometric drawinging of it.

# ISOMETRICS

PROBS 1 & 3: MAKE DOUBLE SIZE ISOMETRIC SKETCHES OF THE OBJECTS.

PROBS 2 & 4: MAKE DOUBLE SIZE INSTRUMENT DRAWINGS OF THE OBJECTS.

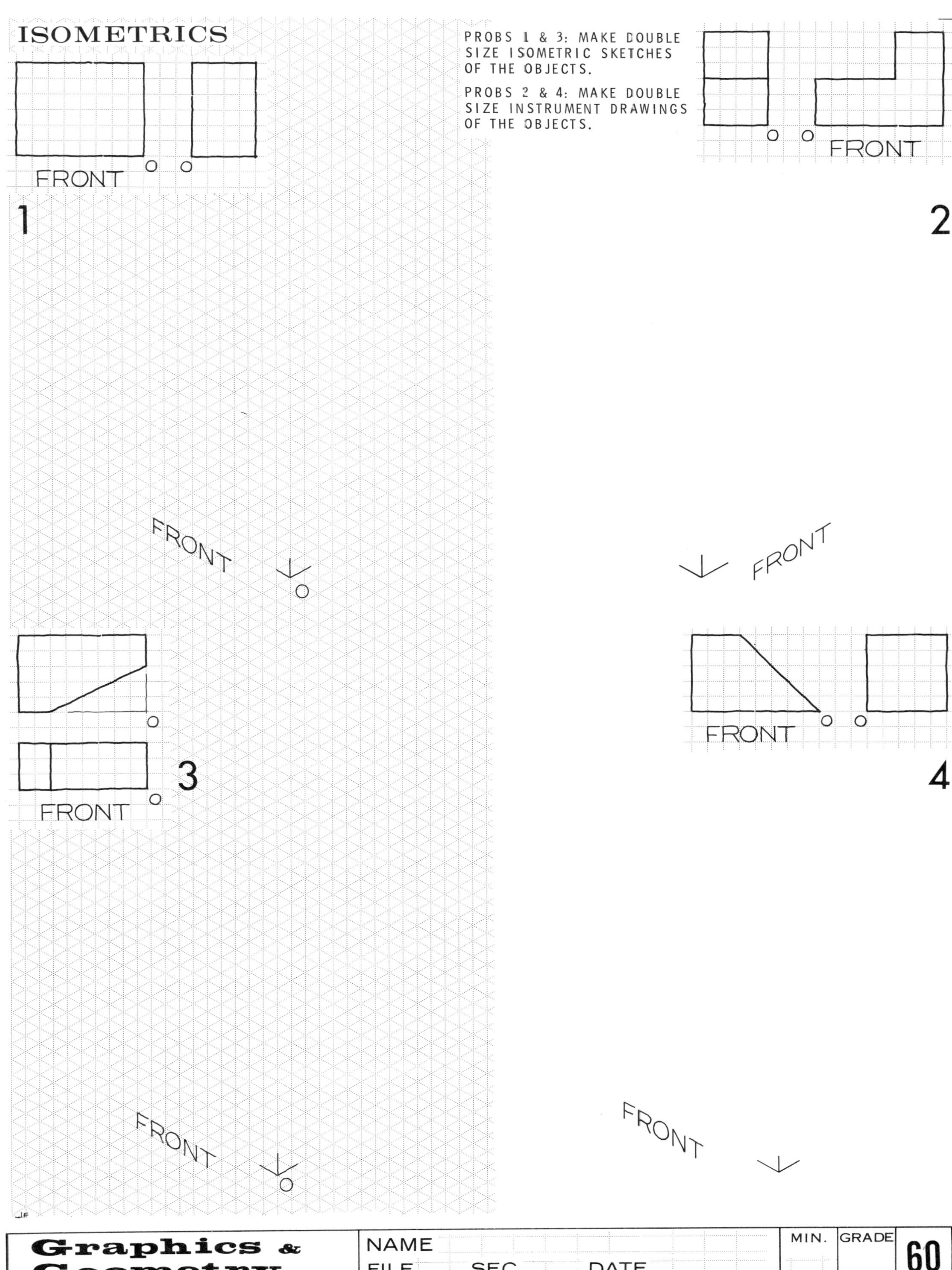

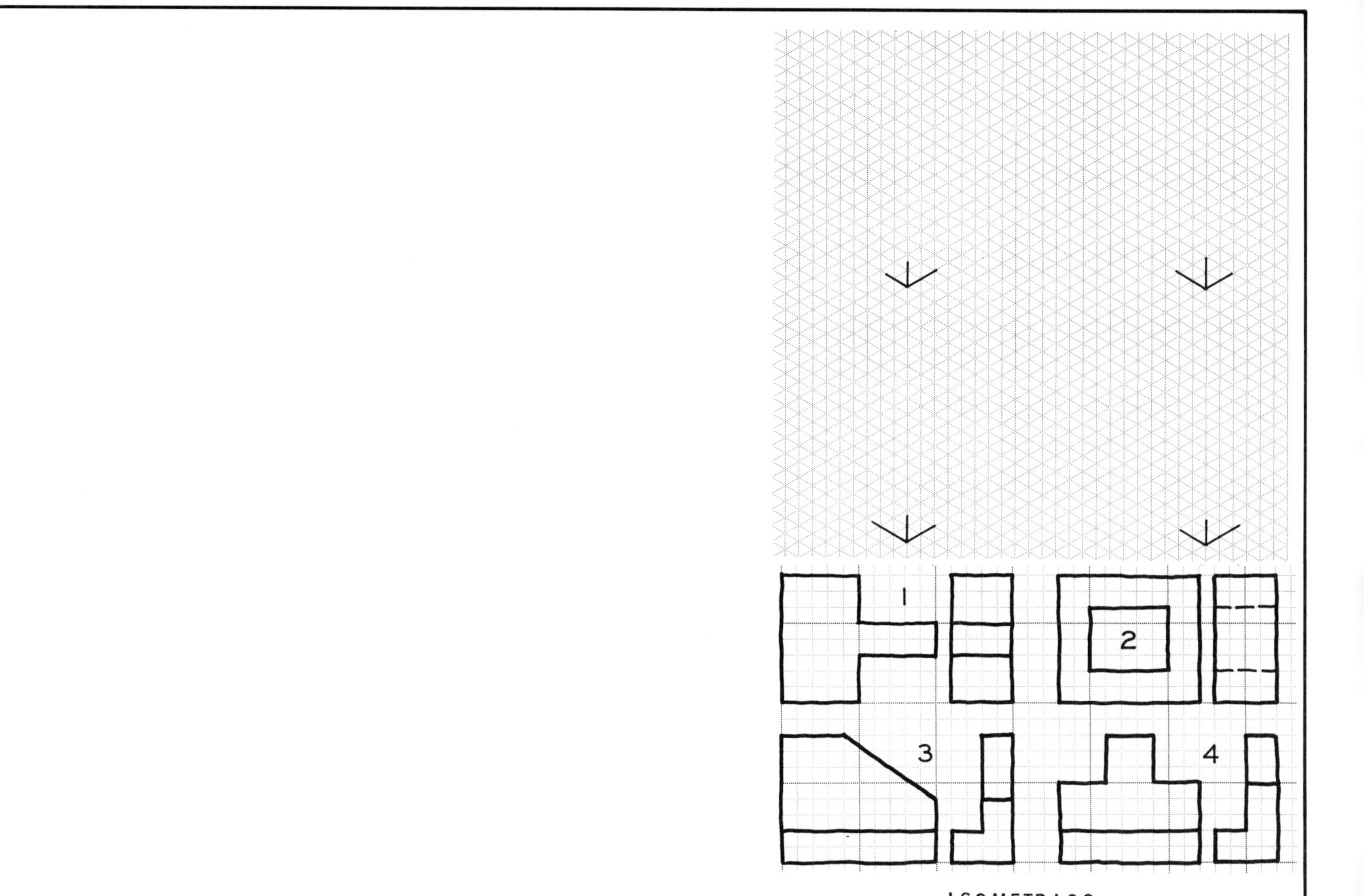


## ISOMETRICS

Using the orthographic views, construct isometric drawings of the parts in the areas indicated above. Or, draw one double-size isometric of the part assigned.

| COMPUTER GRAPHICS | **Graphics for Engineers** © | NAME<br>FILE SEC DATE | MIN | GRADE | 60 |
|---|---|---|---|---|---|

## ISOMETRIC DRAWINGS

**1** DRAW THREE 1" (25 MM) DIA CYLINDERS WITH YOUR ELLIPSE TEMPLATE OR COMPASS USING THE AXES GIVEN. THE VISIBLE ENDS ARE MARKED WITH A "V".

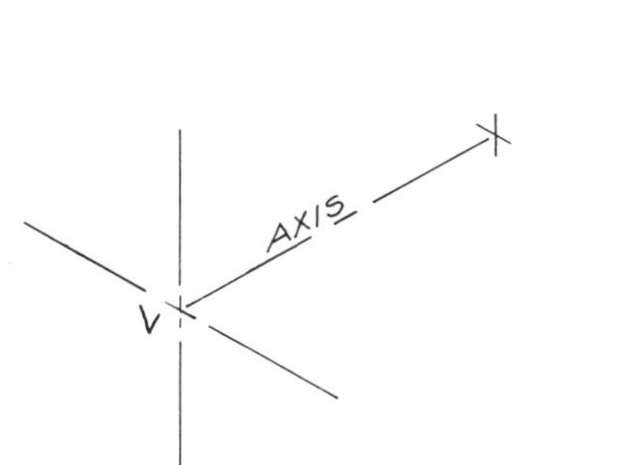


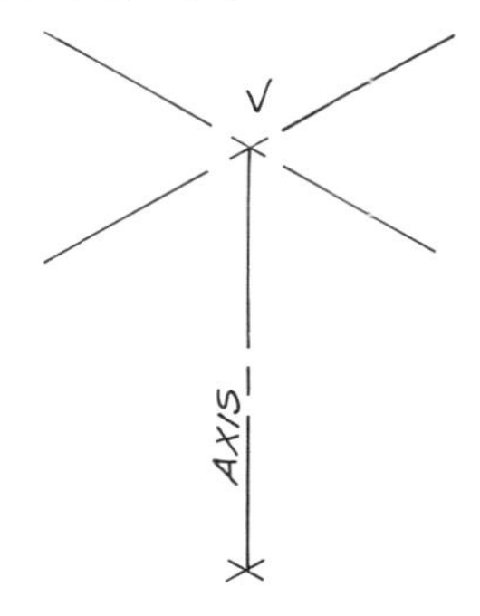


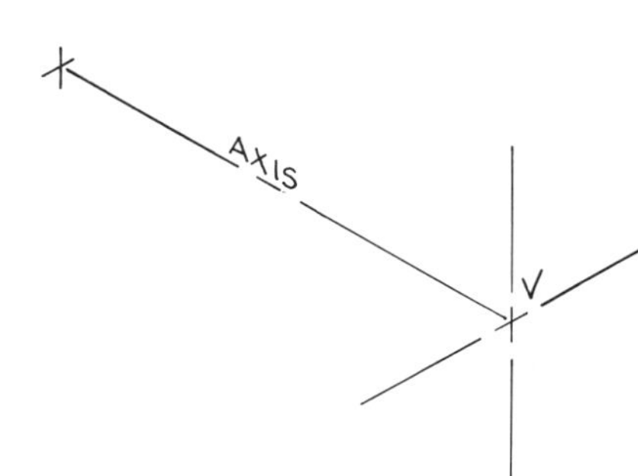


**2** DRAW FULL SIZE ISOMETRICS OF PROBLEMS 2 AND 3 USING YOUR ISOMETRIC TEMPLATE.

**3**

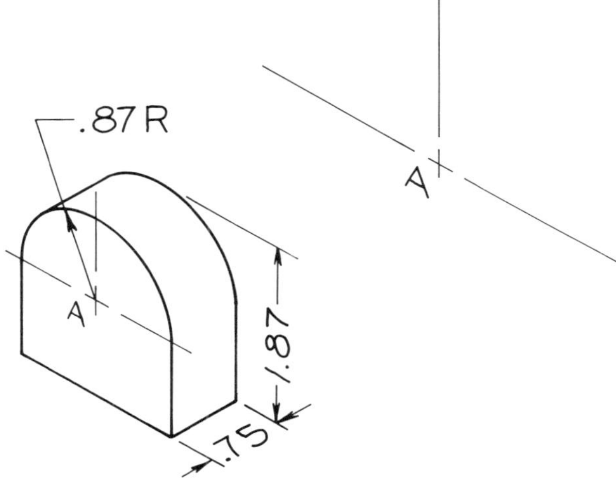


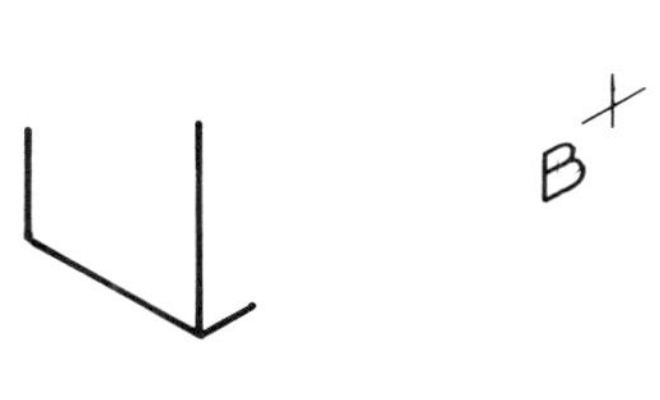

B

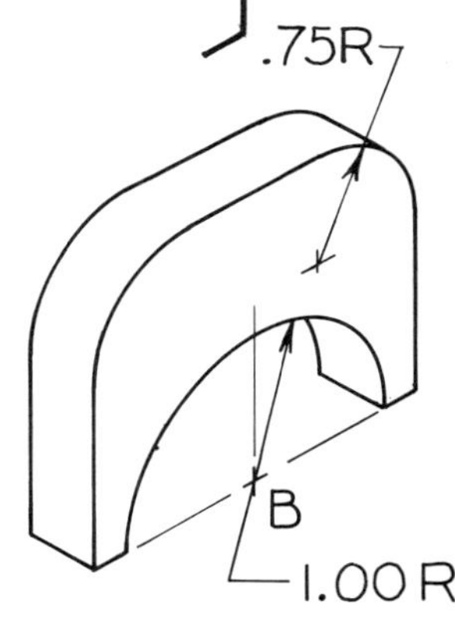


C

**4**

USING FOUR-CENTER CONSTRUCTION, MAKE AN ISOMETRIC DRAWING OF A 4" (100 MM) DIA CYLINDER THAT IS .5" (12 MM) THICK. BEGIN AT POINT C.

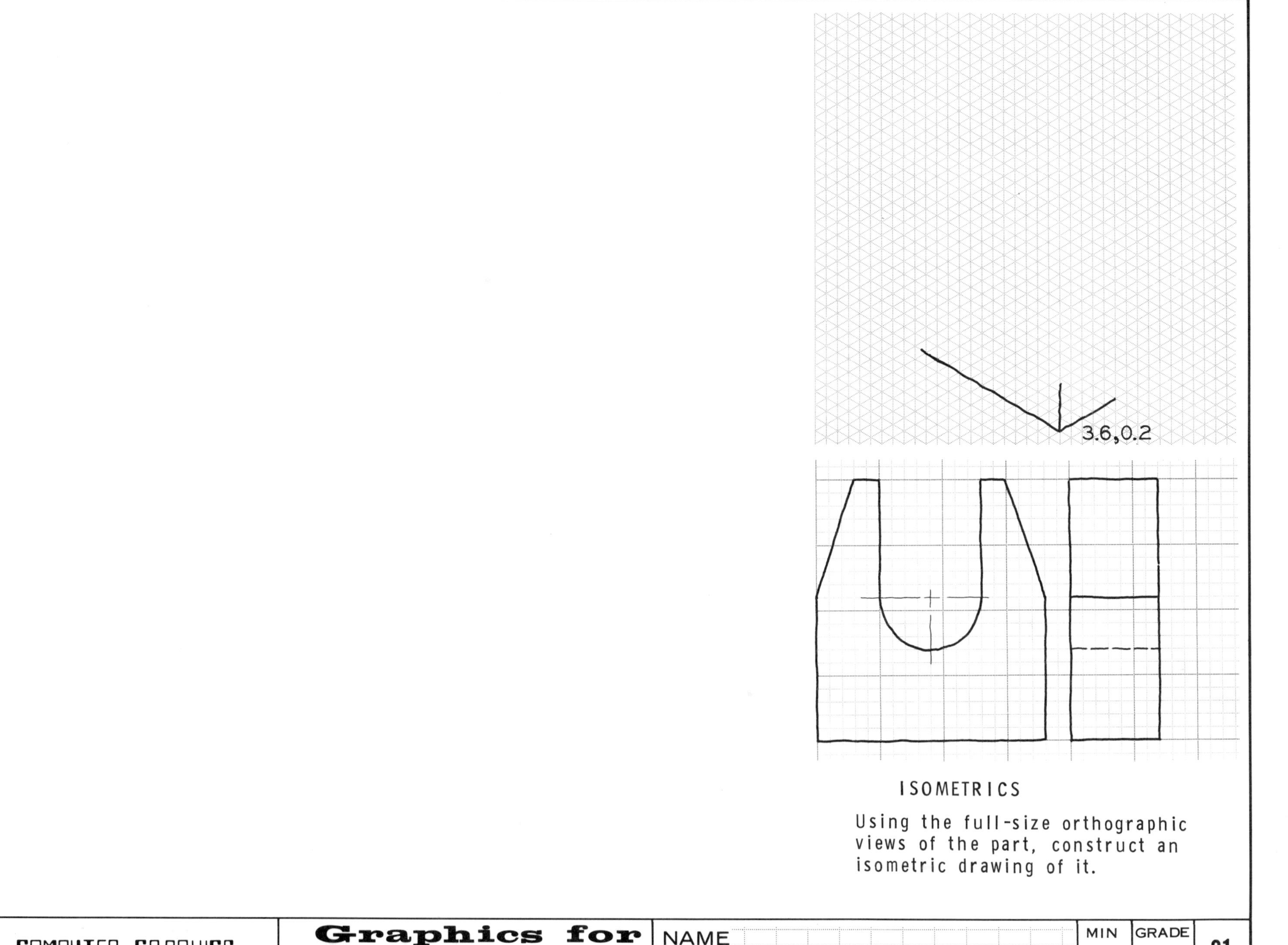


ISOMETRICS

Using the full-size orthographic views of the part, construct an isometric drawing of it.

| COMPUTER GRAPHICS | **Graphics for Engineers** © | NAME<br>FILE SEC DATE | MIN | GRADE | 61 |
|---|---|---|---|---|---|

1 USE ELLIPSE GUIDE OR SKETCH AN ISOMETRIC HOLE OF 1 3/4" DIA IN THE CENTER OF THE SPACER BLANK.

2 MAKE A DOUBLE SIZE ISOMETRIC INSTRUMENT DRAWING OF THE ROLLER BEARING RING. BEGIN AT POINT O.

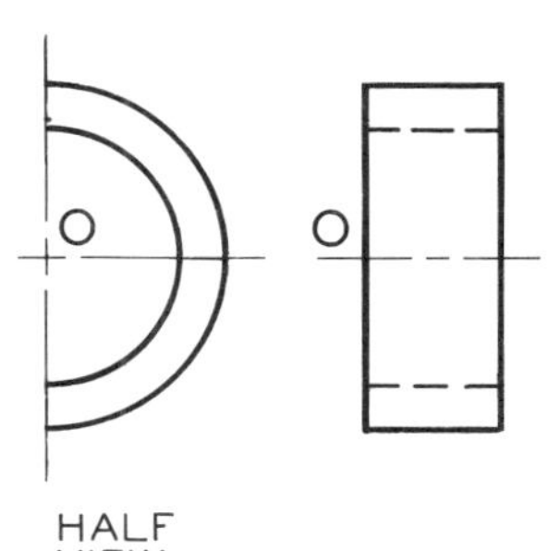

COURTESY OF THE TORRINGTON COMPANY

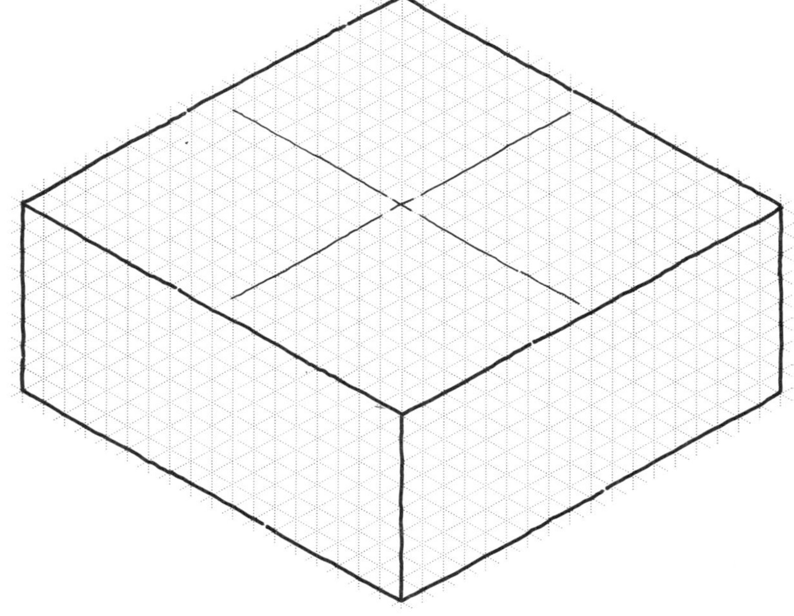

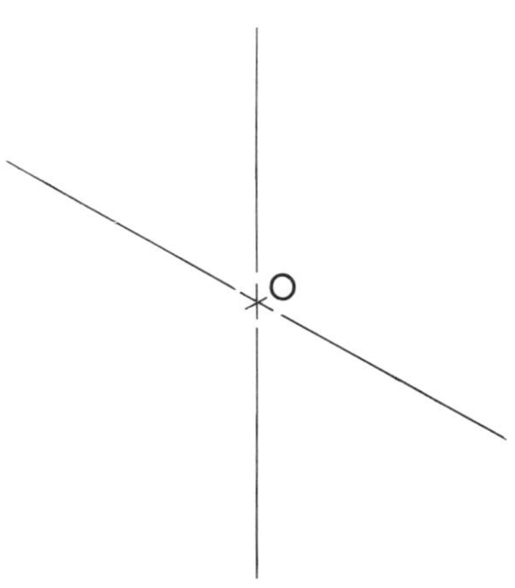


3

CONSTRUCT A FULL SIZE ISOMETRIC DRAWING OF THE RIGHT TOOL. BEGIN AT POINT O.

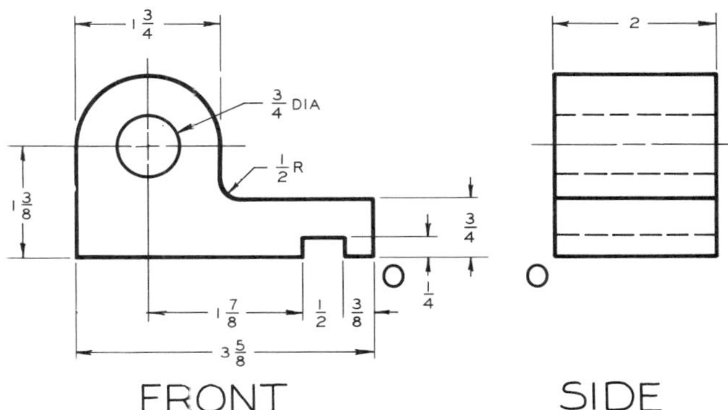


RIGHT TOOL

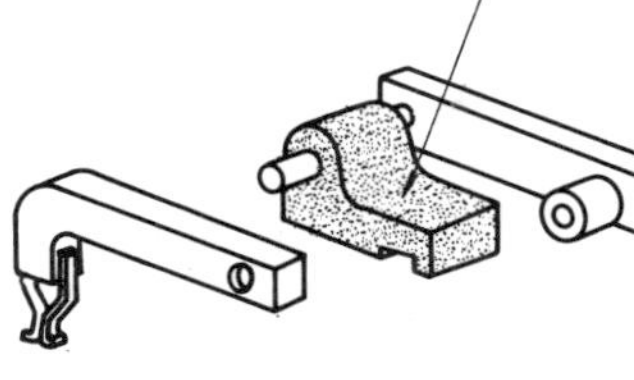

FRONT

O

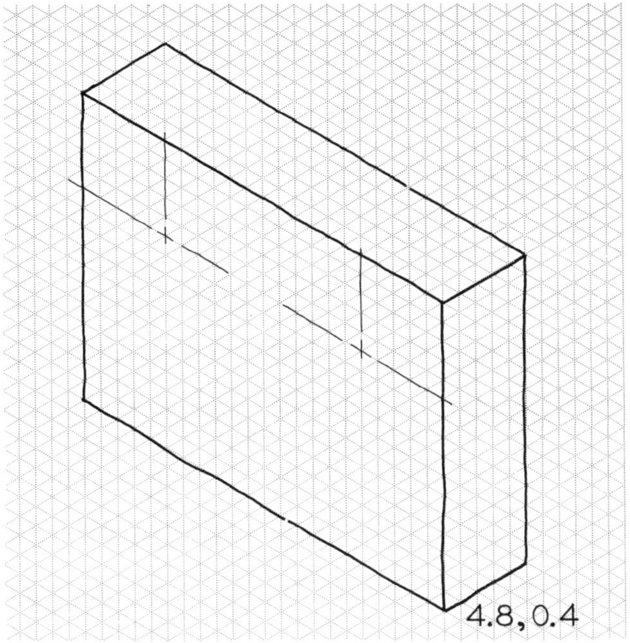

4.8,0.4

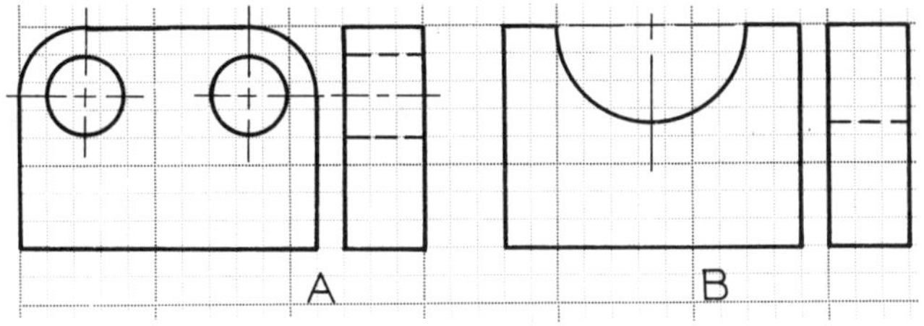

A

B

ISOMETRIC DRAWING

Plot an isometric drawing of the assigned part shown in the half-size views at A and B. Begin the isometric at the corner point given above.

# PIPE SYMBOLS

PROB 1-4: MAKE DOUBLE-SIZE ISOMETRIC DRAWINGS OF THE PIPING SEGMENTS.

PROB 5: DRAW THE RIGHT-SIDE VIEW OF THE PIPING SEGMENT.

PROB 6: MAKE A FULL-SIZE ISOMETRIC DRAWING OF PROB 5.

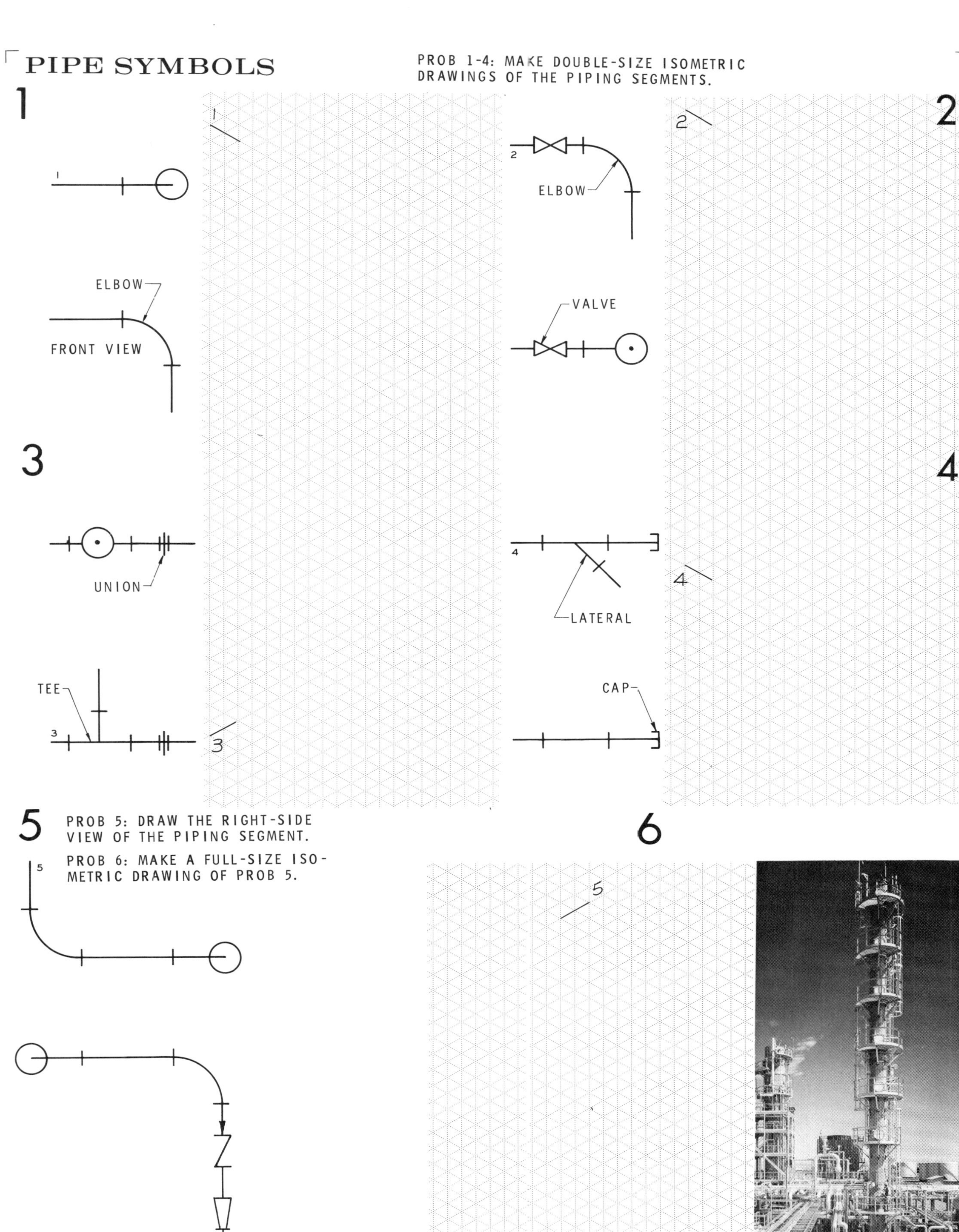

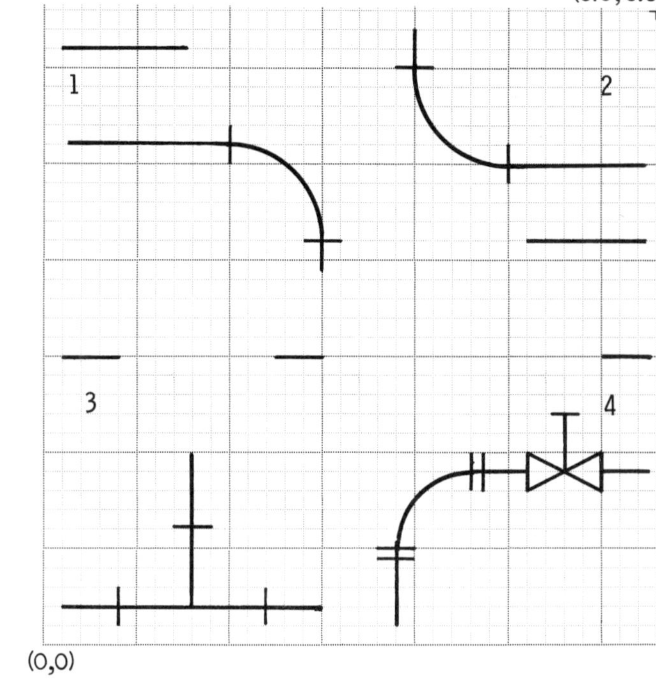


PIPING SYMBOLS

1. Complete the orthographic views of the piping connections.
2. Construct isometric drawings of the piping connections on a separate sheet.

**1** DRAW THE MISSING VIEWS OF EACH LINE AND INDICATE WHAT TYPE OF LINE EACH IS. LABEL THE TRUE LENGTH VIEWS TL.

TYPE: 

ABBREVIATIONS

HL = HORIZ. LINE
FL = FRONTAL LINE
PL = PROFILE LINE
OL = OBLIQUE LINE

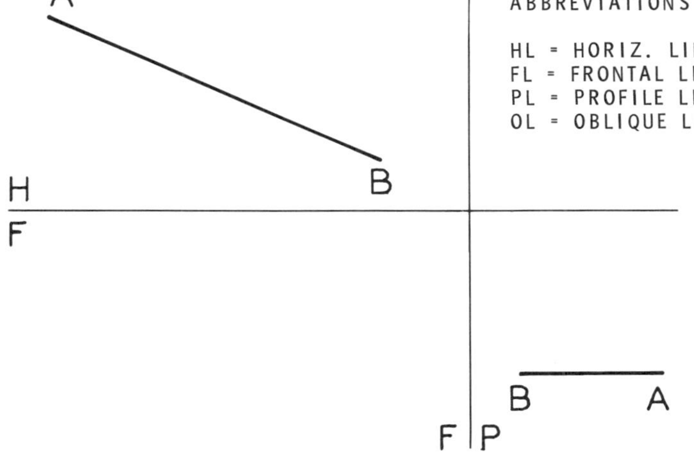

**2** TYPE: 

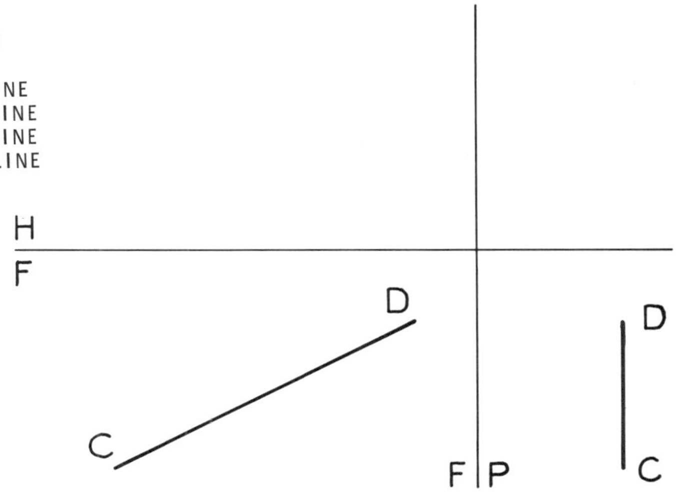

**3** TYPE: 

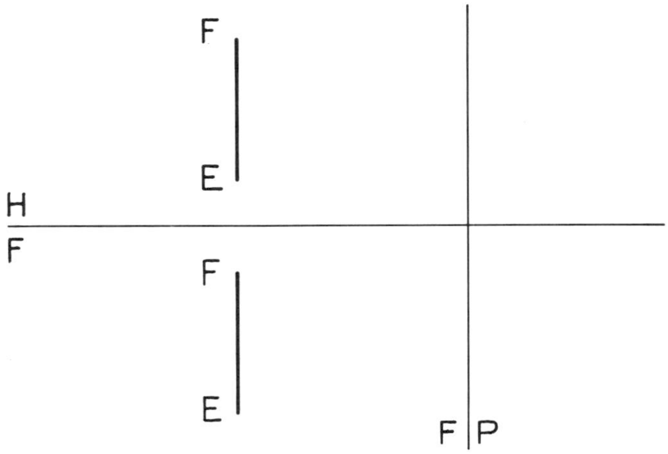

**4** TYPE: 

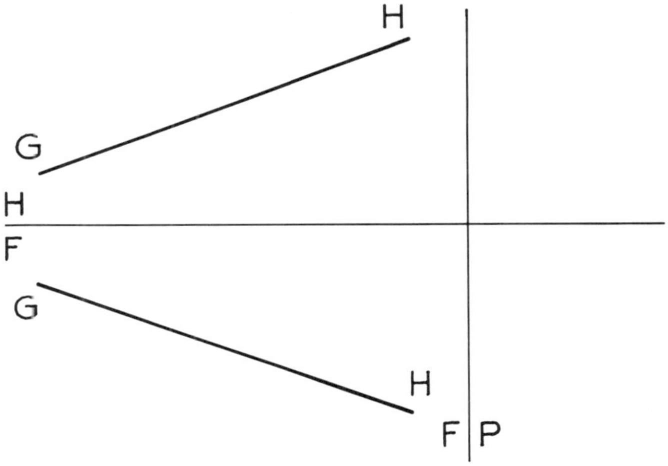

**5** TYPE: 

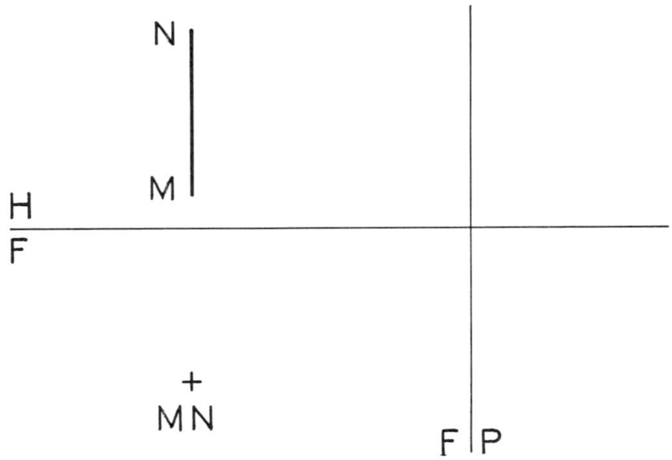

**6** TYPE:

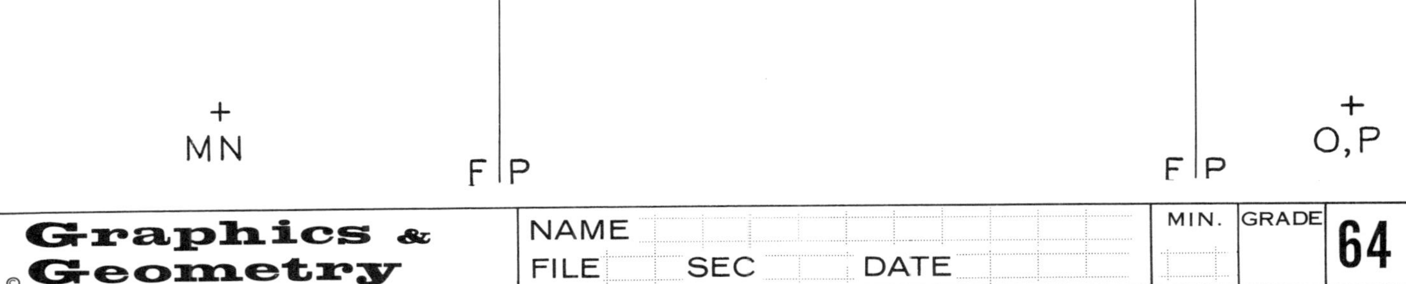

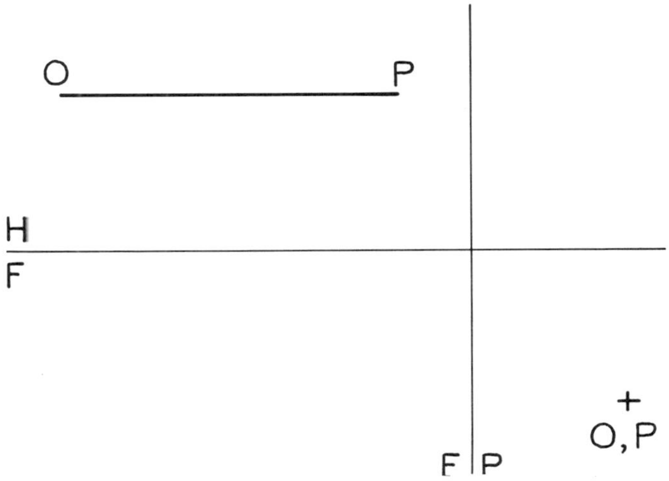

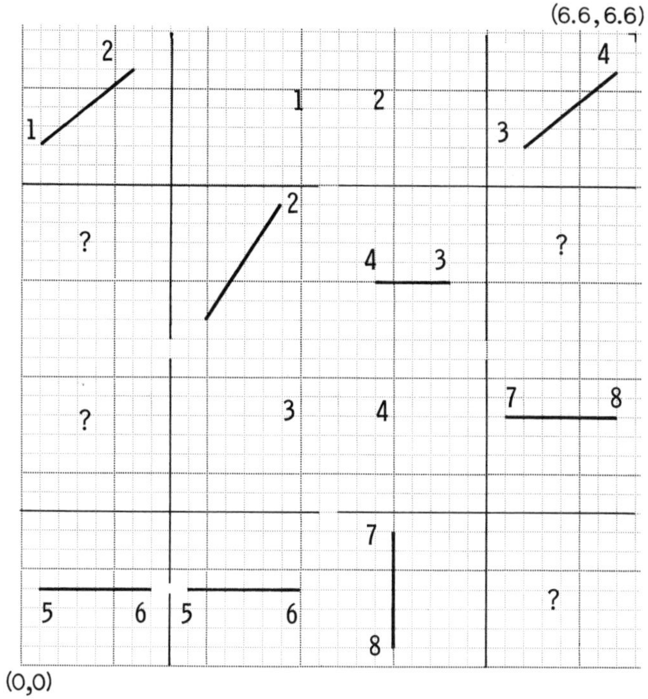

(6.6,6.6)

(0,0)

LINES IN ORTHOGRAPHIC

Plot the given views of the lines along with their reference lines. Find the missing views, label the lines, and label the reference lines.

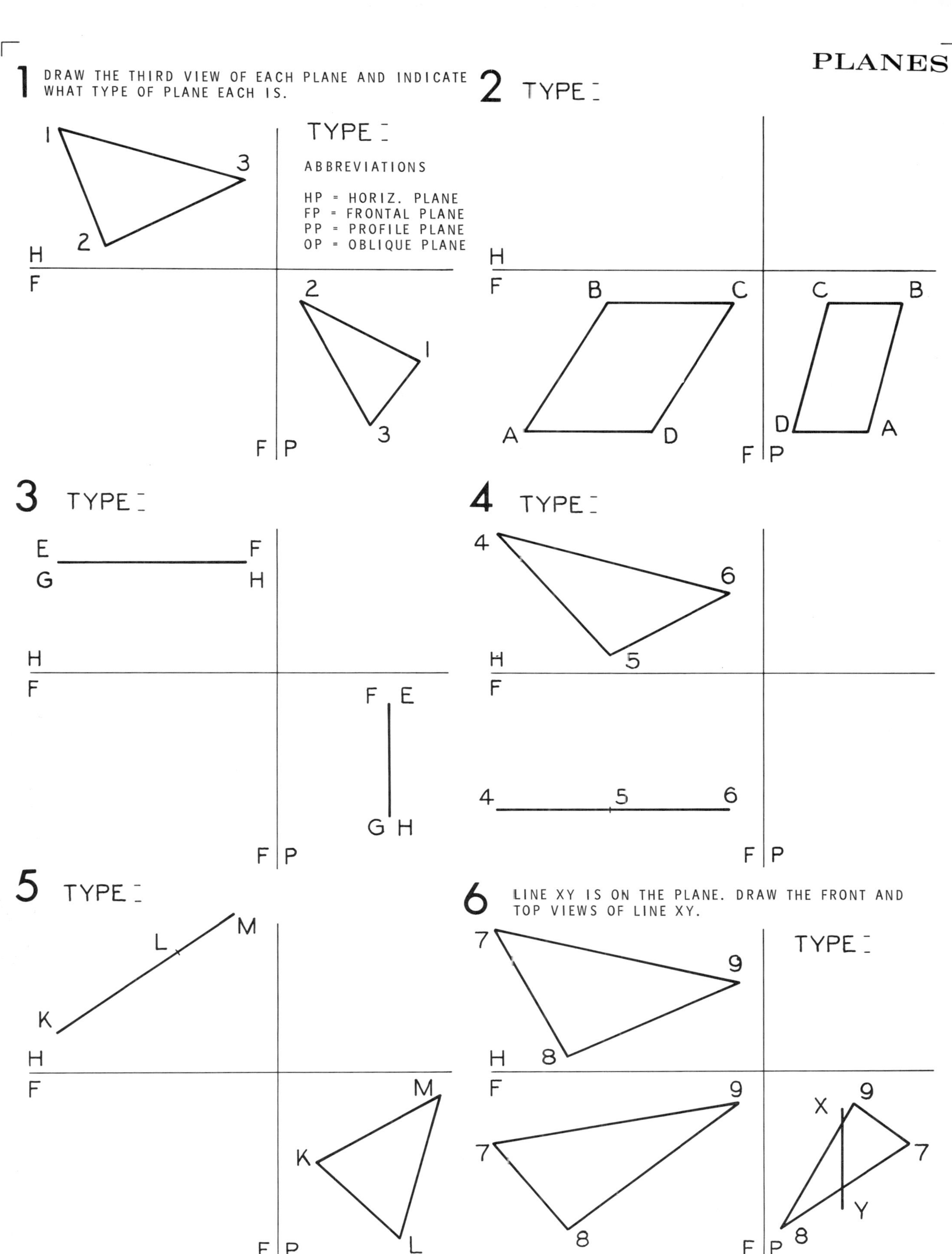
1 DRAW THE THIRD VIEW OF EACH PLANE AND INDICATE WHAT TYPE OF PLANE EACH IS.
TYPE
ABBREVIATIONS
HP = HORIZ. PLANE
FP = FRONTAL PLANE
PP = PROFILE PLANE
OP = OBLIQUE PLANE
1
2
3
H
F
P
2 TYPE
A
B
C
D
3 TYPE
E
F
G
H
4 TYPE
4
5
6
5 TYPE
K
L
M
6 LINE XY IS ON THE PLANE. DRAW THE FRONT AND TOP VIEWS OF LINE XY.
TYPE
7
8
9
X
Y

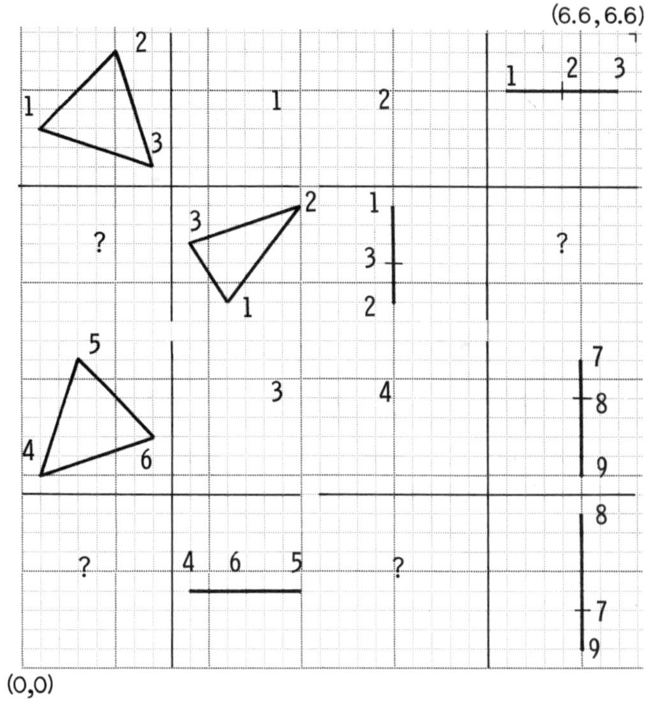

(6.6,6.6)

(0,0)

## ORTHOGRAPHIC PROJECTION

Plot the given views of the planes and find the missing views. Label all points and label the reference lines. Identify the planes as oblique (OP), frontal (FP), horizontal (HP), or profile (PP).

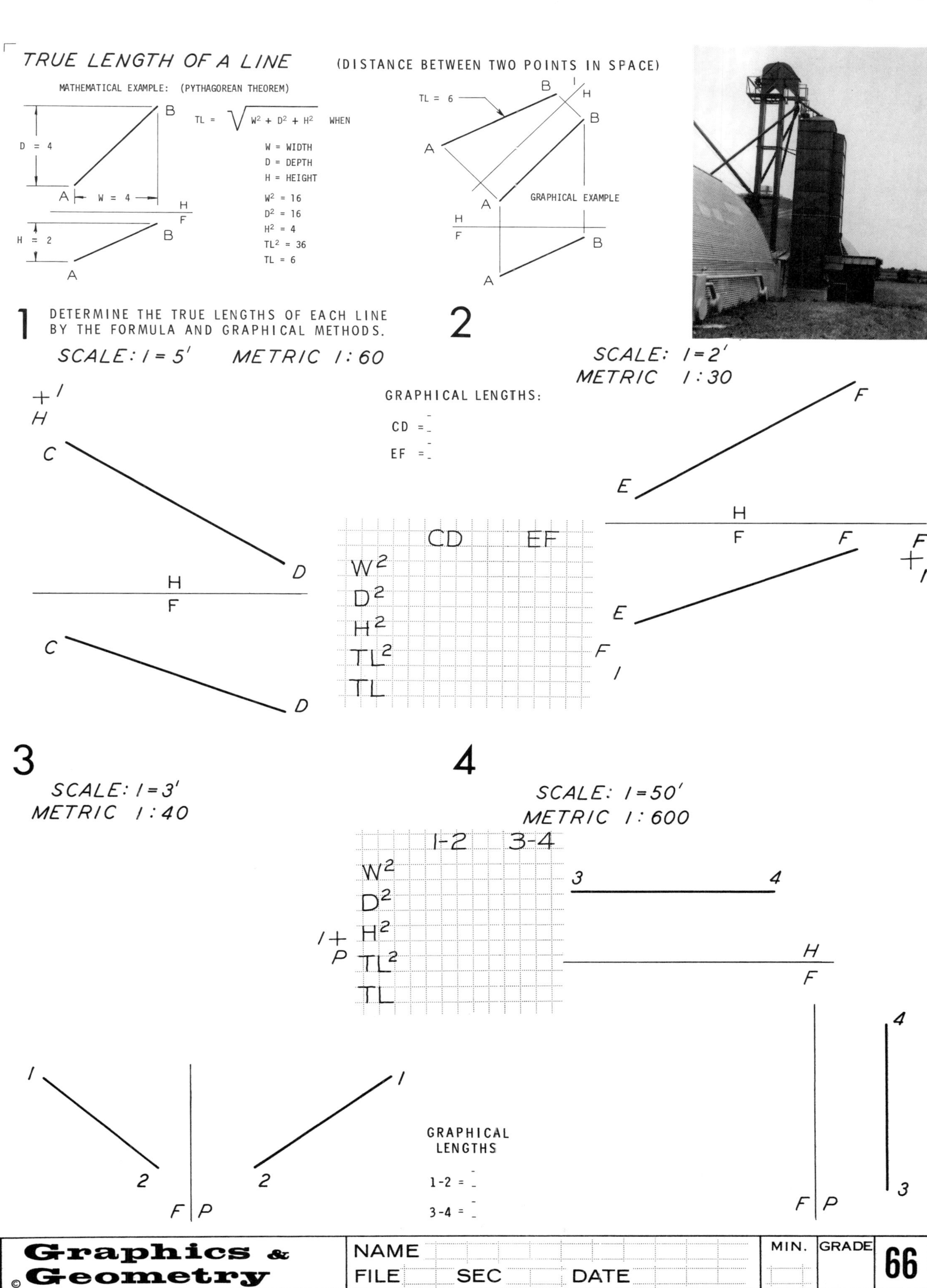

TRUE LENGTH OF A LINE
(DISTANCE BETWEEN TWO POINTS IN SPACE)
MATHEMATICAL EXAMPLE: (PYTHAGOREAN THEOREM)
TL = $\sqrt{W^2 + D^2 + H^2}$ WHEN
W = WIDTH
D = DEPTH
H = HEIGHT
$W^2$ = 16
$D^2$ = 16
$H^2$ = 4
$TL^2$ = 36
TL = 6
D = 4
W = 4
H = 2
TL = 6
GRAPHICAL EXAMPLE
1
DETERMINE THE TRUE LENGTHS OF EACH LINE BY THE FORMULA AND GRAPHICAL METHODS.
SCALE: 1 = 5′ METRIC 1:60
2
SCALE: 1=2′
METRIC 1:30
GRAPHICAL LENGTHS:
CD =
EF =
CD EF
$W^2$
$D^2$
$H^2$
$TL^2$
TL
3
SCALE: 1=3′
METRIC 1:40
4
SCALE: 1=50′
METRIC 1:600
1-2 3-4
$W^2$
$D^2$
$H^2$
$TL^2$
TL
GRAPHICAL LENGTHS
1-2 =
3-4 =
Graphics & Geometry
NAME
FILE SEC DATE
MIN. GRADE
66

# LETTERING GUIDE LINES

THESE GUIDELINES CAN BE USED TO UNDERLAY TRACING PAPER OR OTHER TRANSLUSCENT PAPERS WHEN LETTERING A DRAWING.

| **Graphics for Engineers** © | NAME<br>FILE SEC DATE | MIN. | GRADE | |
|---|---|---|---|---|

## TRUE LENGTH OF A LINE — APPLICATION

BELOW ARE THE PARTIAL FRONT AND TOP VIEWS OF AN ENGINE MOUNT FOR A HELICOPTER. DETERMINE THE TRUE LENGTHS OF THE CENTER LINES OF THE MEMBERS LISTED IN THE TABLE BELOW. USE THE FOLLOWING METHODS:

1. BY AUXILIARY VIEWS
2. BY A TRUE-LENGTH DIAGRAM

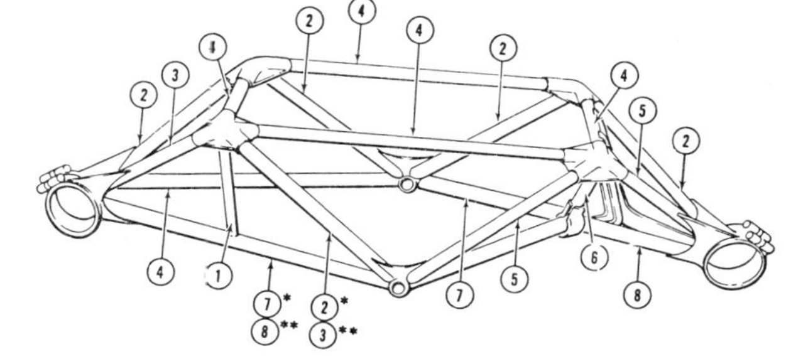

| | TL |
|---|---|
| AB | |
| BC | |
| CD | |
| CE | |

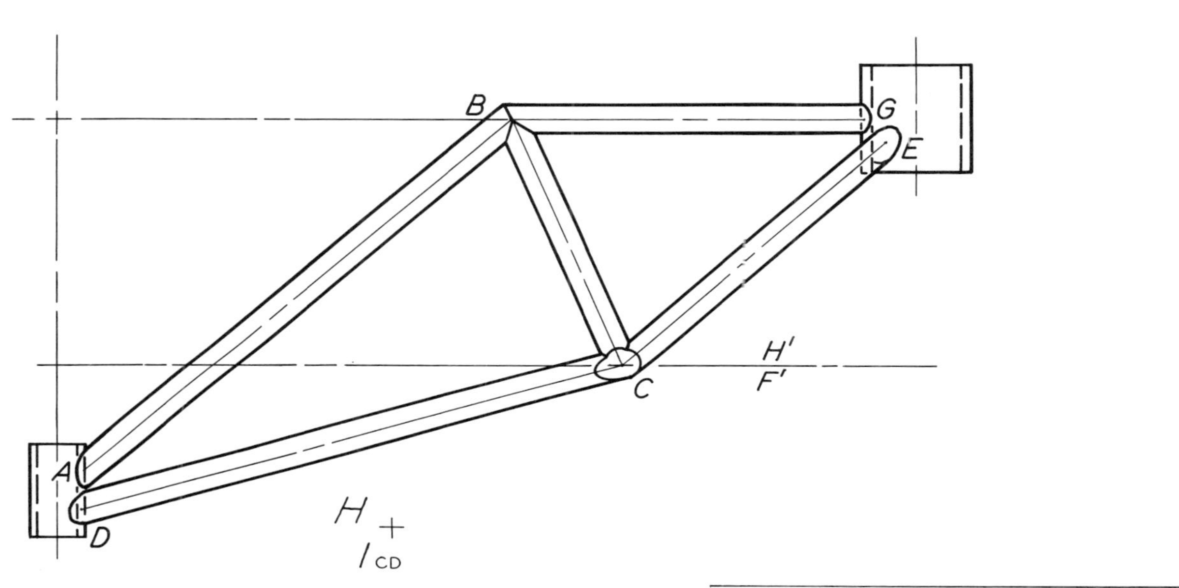


TL DIAGRAM

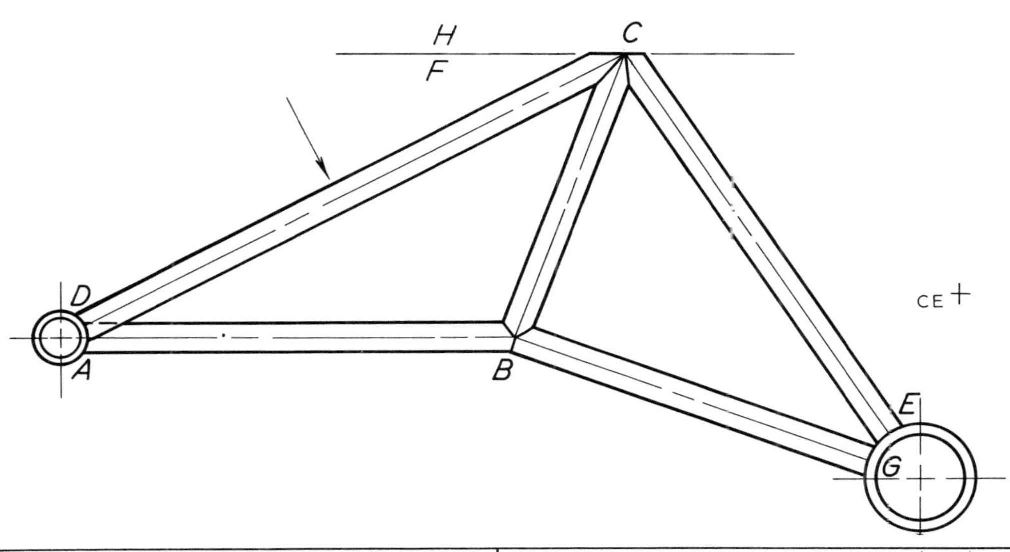


SCALE: 1 = 4'
METRIC: 1:50

| Graphics & Geometry | NAME | MIN. | GRADE | 67 |
|---|---|---|---|---|
| © | FILE SEC DATE | | | |

# LETTERING GUIDE LINES

THESE GUIDELINES CAN BE USED TO UNDERLAY TRACING PAPER OR OTHER TRANSLUSCENT PAPERS WHEN LETTERING A DRAWING.

| Graphics for Engineers © | NAME<br>FILE SEC DATE | MIN. | GRADE | |
|---|---|---|---|---|

# SLOPE AND PERCENT GRADE

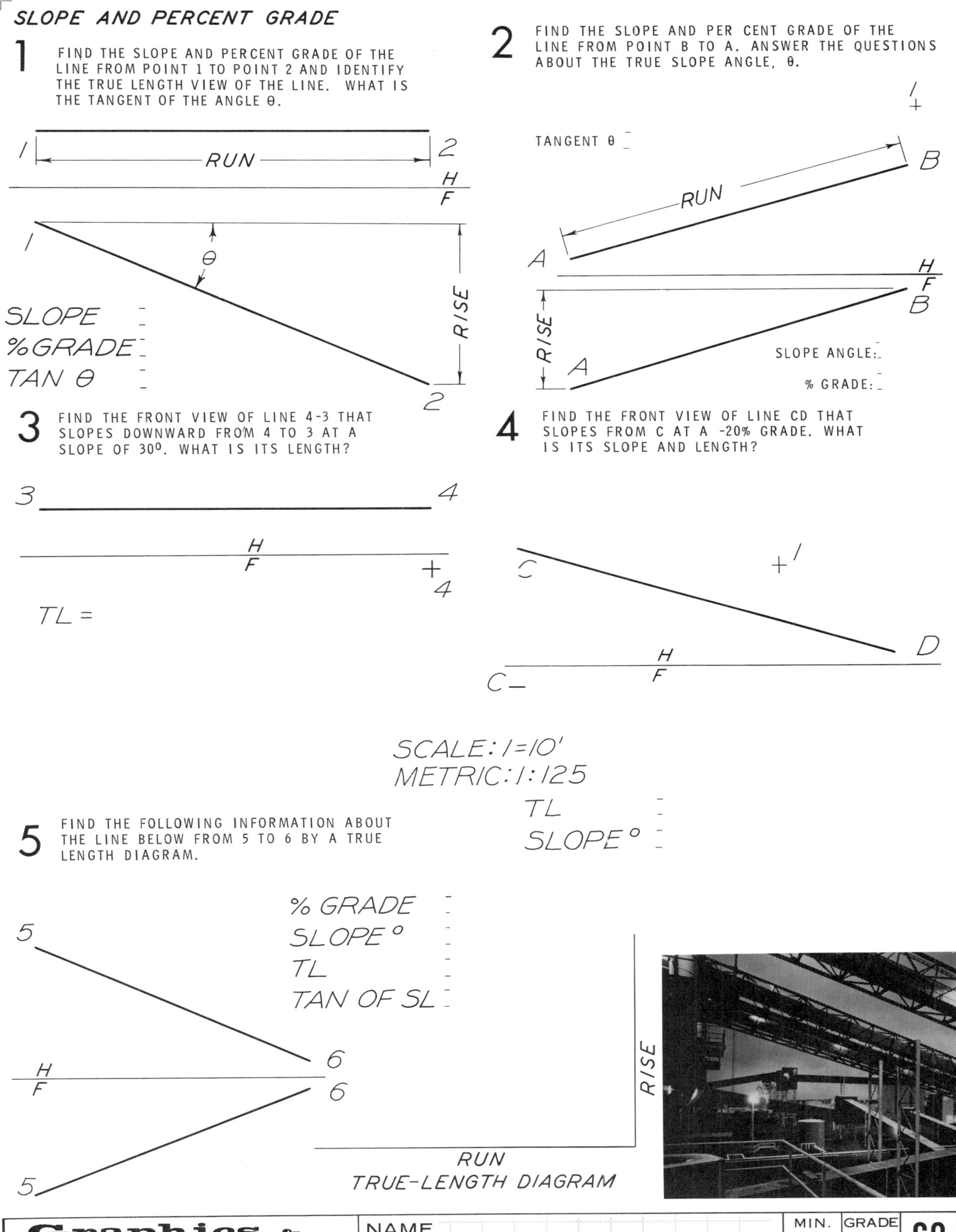

## SLOPE AND BEARING OF LINES

**1** DETERMINE THE COMPASS BEARINGS OF LINES 0-1 AND 0-3, AND THE AZIMUTH BEARINGS OF LINES 0-2 AND 0-4. LETTER YOUR ANSWERS ALONG EACH LINE.

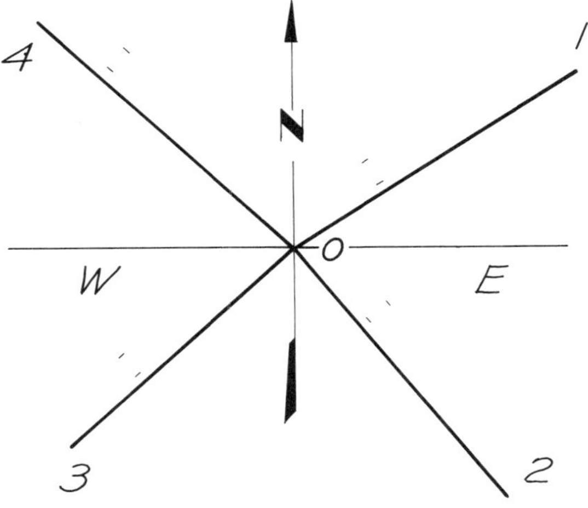


**2** DETERMINE THE SLOPE AND BEARING OF LINE AB. EXPRESS THE SLOPE IN % GRADE AND DEGREES. INDICATE BEARING TOWARDS THE LOW END.

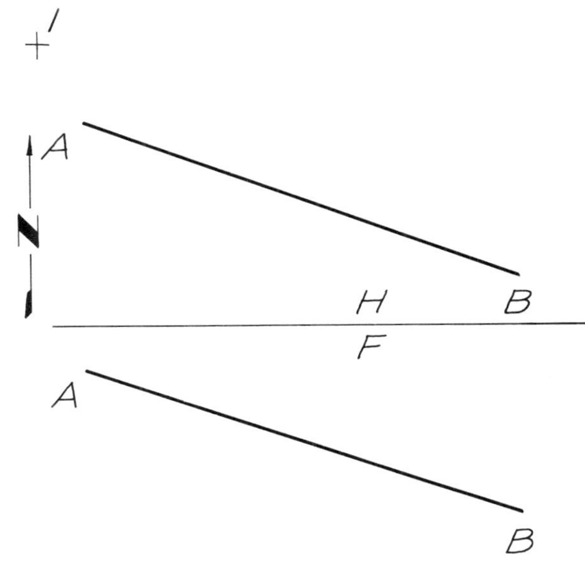


BEARING:
SLOPE °:
% GRADE:

**3** FIND THE BEARING, SLOPE AND TRUE LENGTH OF THE LONG RANGE TOWER EXCAVATOR TRACK CABLE 0-1 THAT LEADS FROM THE TOP OF THE STRUCTURE. CONSTRUCT A SECOND CABLE 0-2 THAT BEARS DUE EAST WITH A -70% GRADE. COMPLETE THE TABLE BELOW.

| CABLE | TL | BEAR | SL |
|---|---|---|---|
| O-1 | | | |
| O-2 | | | |

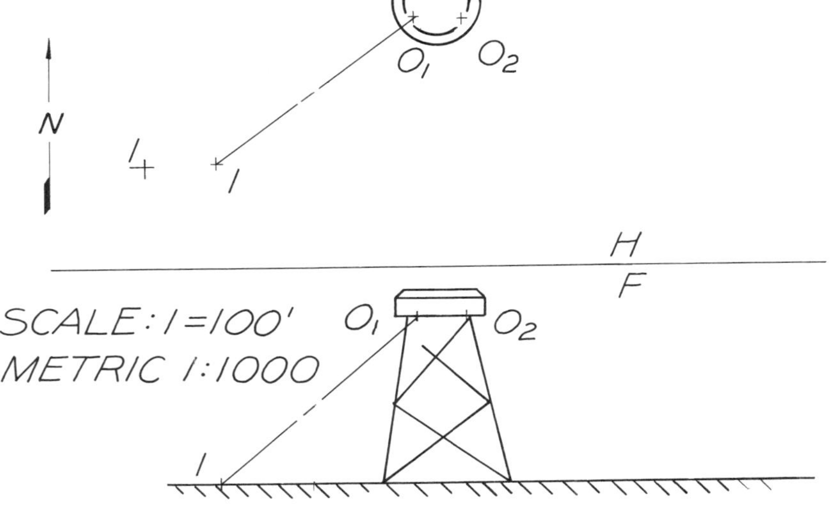

Courtesy of Sauerman Bros. Inc.

| Graphics & Geometry | NAME | MIN. | GRADE | 69 |
|---|---|---|---|---|
| © | FILE SEC DATE | | | |

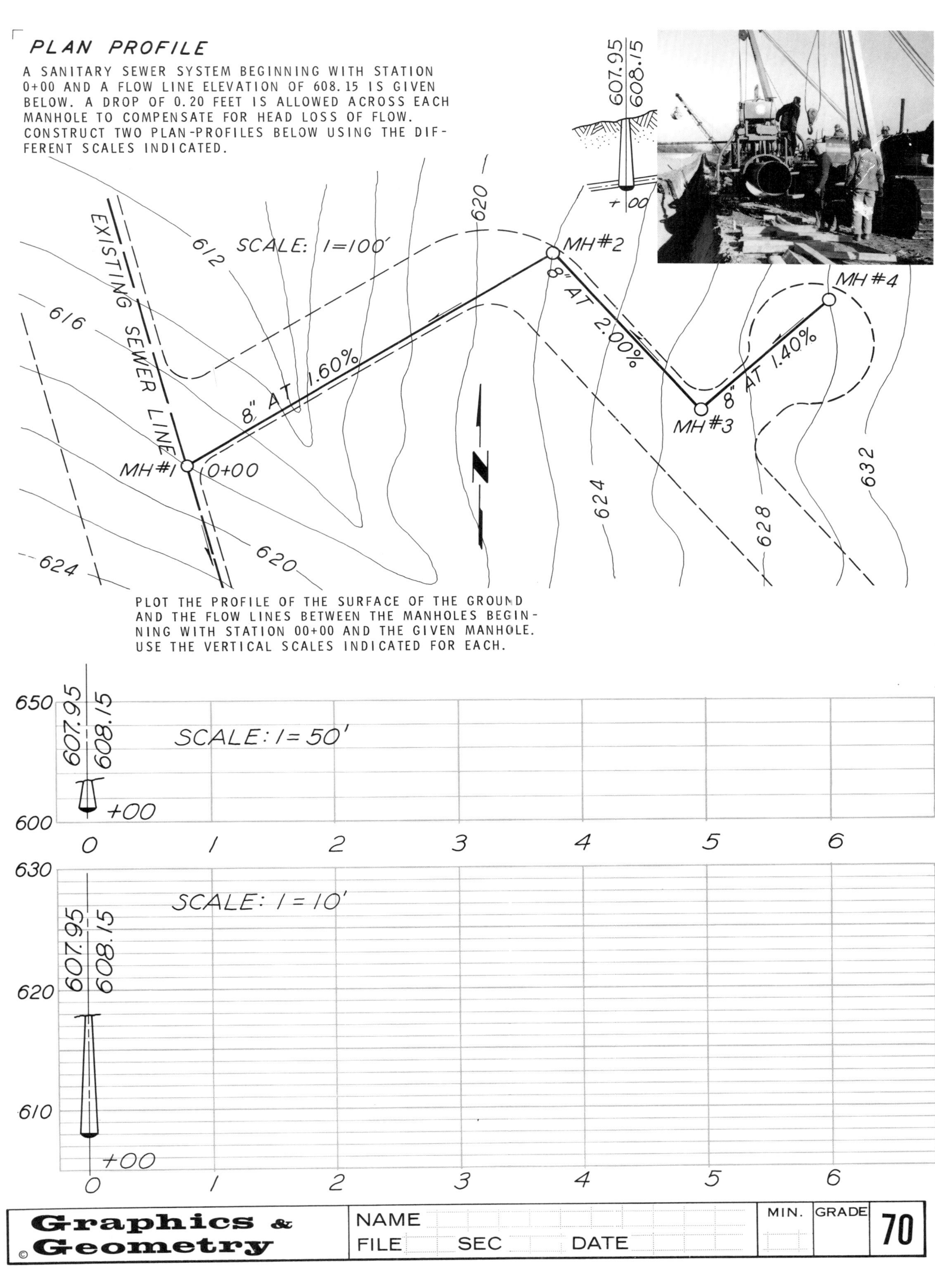
PLAN PROFILE
A SANITARY SEWER SYSTEM BEGINNING WITH STATION 0+00 AND A FLOW LINE ELEVATION OF 608.15 IS GIVEN BELOW. A DROP OF 0.20 FEET IS ALLOWED ACROSS EACH MANHOLE TO COMPENSATE FOR HEAD LOSS OF FLOW. CONSTRUCT TWO PLAN-PROFILES BELOW USING THE DIFFERENT SCALES INDICATED.
607.95
608.15
+00
SCALE: 1=100'
EXISTING SEWER LINE
MH#1
0+00
8" AT 1.60%
MH#2
8" AT 2.00%
MH#3
8" AT 1.40%
MH #4
N
612
616
620
620
624
624
628
632
PLOT THE PROFILE OF THE SURFACE OF THE GROUND AND THE FLOW LINES BETWEEN THE MANHOLES BEGINNING WITH STATION 00+00 AND THE GIVEN MANHOLE. USE THE VERTICAL SCALES INDICATED FOR EACH.
650
600
607.95
608.15
+00
SCALE: 1=50'
0
1
2
3
4
5
6
630
620
610
607.95
608.15
+00
SCALE: 1=10'
0
1
2
3
4
5
6
Graphics & Geometry
©
NAME
FILE
SEC
DATE
MIN.
GRADE
70

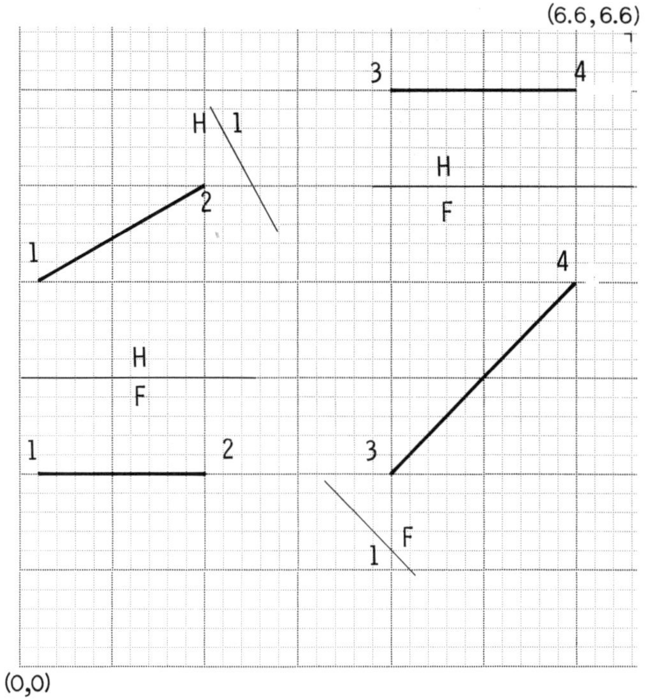

(6.6,6.6)

3 4

H 1

H

F

2

1

4

H

F

1 2

3

F

1

(0,0)

Find the point views of the lines.

PROBLEMS 1 & 2: FIND THE POINT VIEWS OF THE LINES.

PROBLEMS 3 & 4: FIND THE EDGE VIEWS OF THE PLANES.

+1

3 4

# 1

# 2

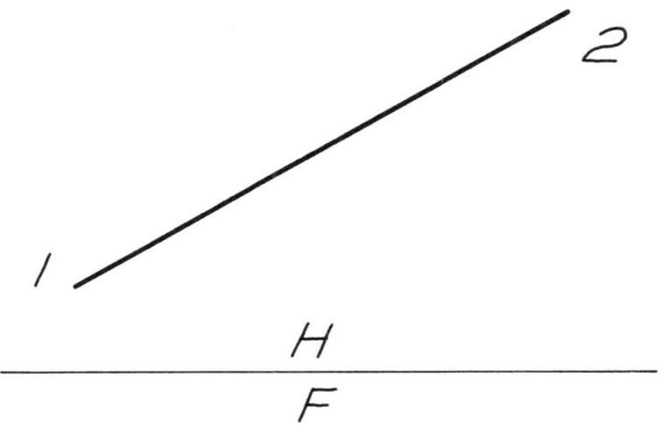


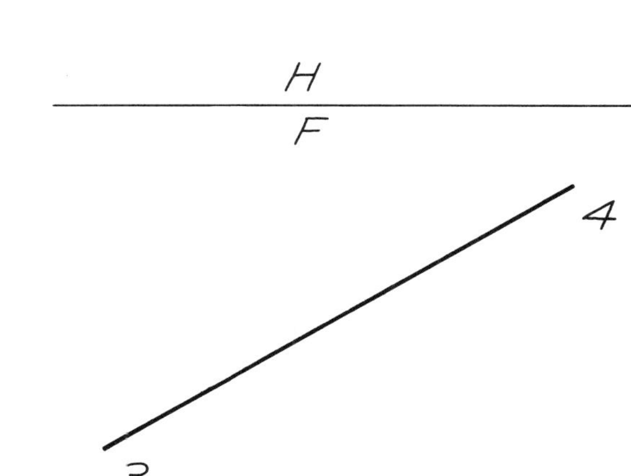


1+

1 2

# 3

# 4

A
B
1+
C

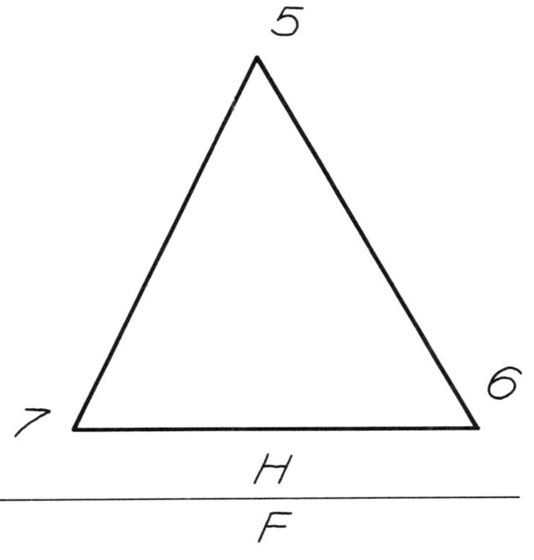


H
F
A

5
+1
B

6
C
7

JAE

| **Graphics & Geometry** © | NAME<br>FILE SEC DATE | MIN. | GRADE | 71 |
|---|---|---|---|---|

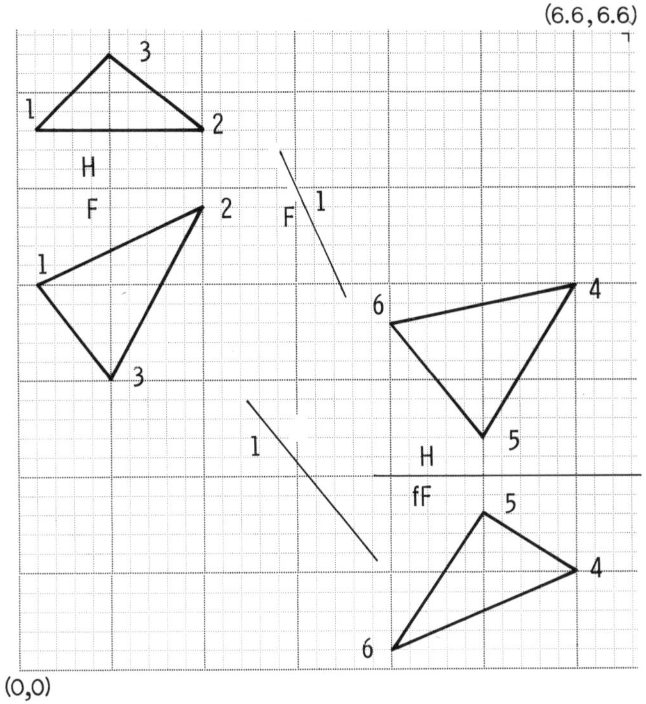

(6.6, 6.6)

3
1 2
H
F
2
1
3
F 1
6 4
5
1
H
fF
5
4
6
(0,0)

Find the edge views of the planes.

# PIERCING POINT OF A LINE ON A PLANE

**1** FIND THE PIERCING POINT OF THE LINE AND PLANE BY AN AUXILIARY VIEW AND SHOW VISIBILITY IN ALL VIEWS.

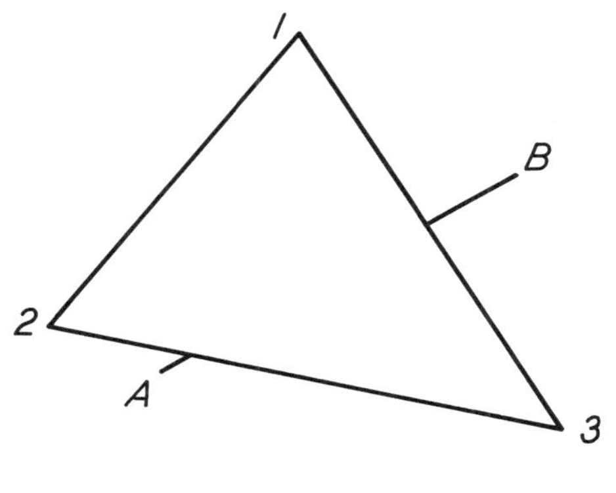

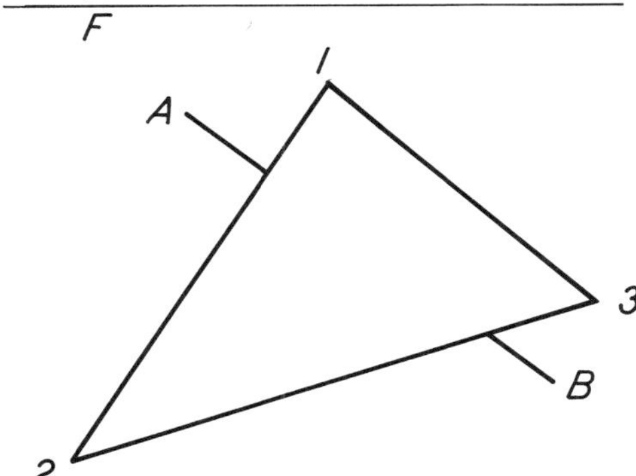


**2** CONSTRUCT A PERPENDICULAR FROM THE POINT TO THE PLANE BY AN AUXILIARY VIEW. SHOW THE LINE IN ALL VIEWS; SHOW VISIBILITY.

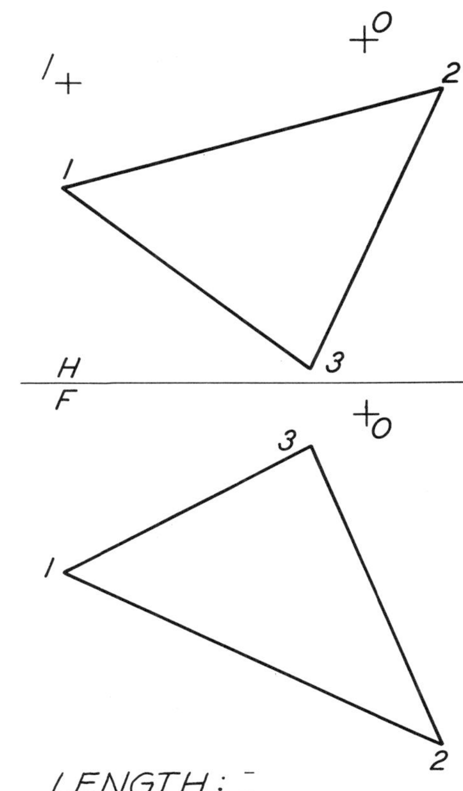


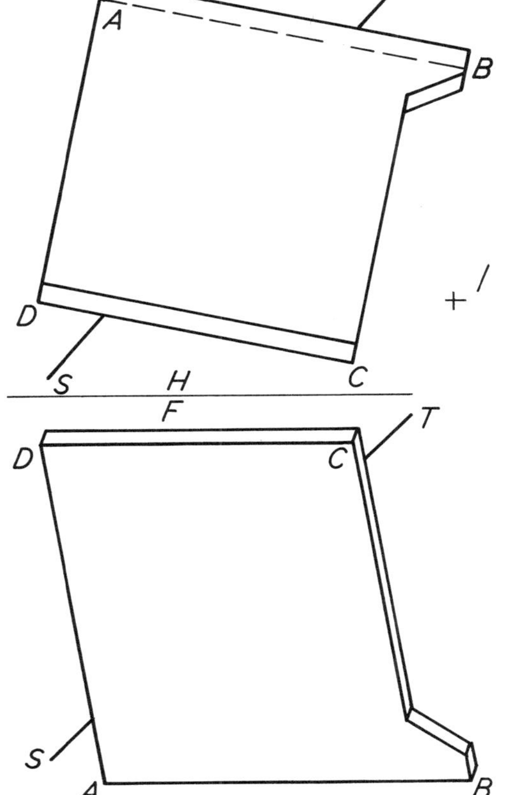


SCALE: 1=3'

SCALE: 1:40

**3**

FIND THE PIERCING POINT OF CENTER LINE ST WITH THE BRACKET PLATE OF AN AIRCRAFT'S LANDING GEAR SYSTEM. SHOW VISIBILITY IN ALL VIEWS.

COURTESY OF NORTH AMERICAN AVIATION

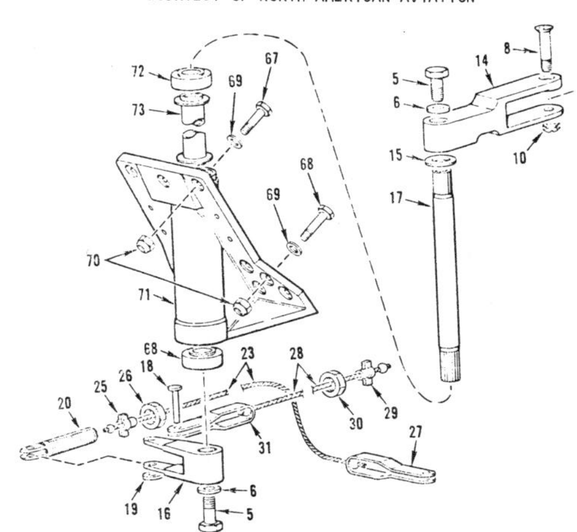

| Graphics & Geometry © | NAME | MIN. | GRADE | 72 |
|---|---|---|---|---|
| | FILE SEC DATE | | | |

# METRIC SCALES

**Remove this page and fold it longwise along the desired scale for making metric measurements.**

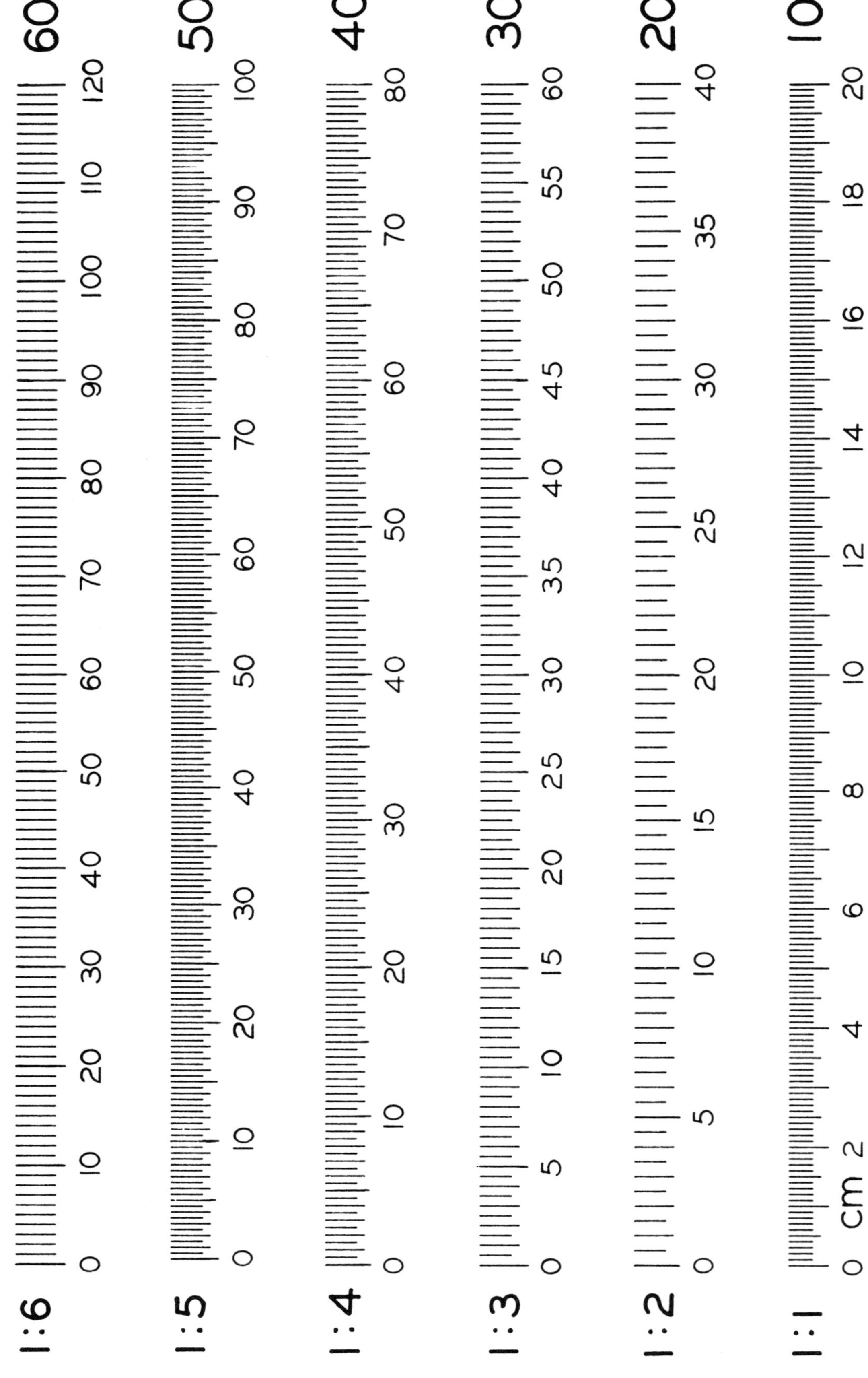

## SLOPE & SLOPE DIRECTION OF PLANES

1 FIND THE SLOPE AND DIRECTION OF SLOPE FOR PLANE ABC.

SLOPE DIRECTION ___

SLOPE ANGLE ___

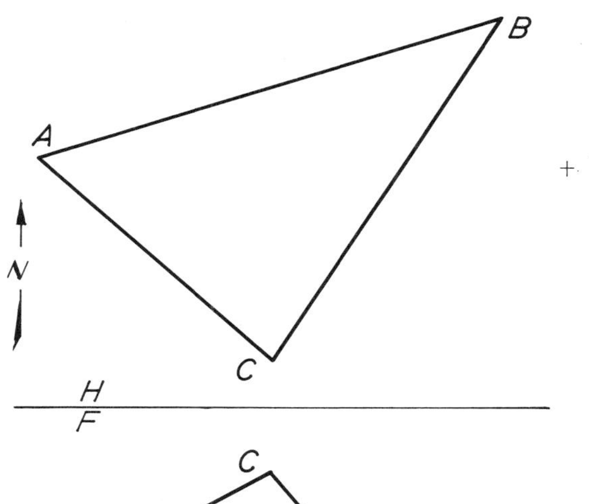


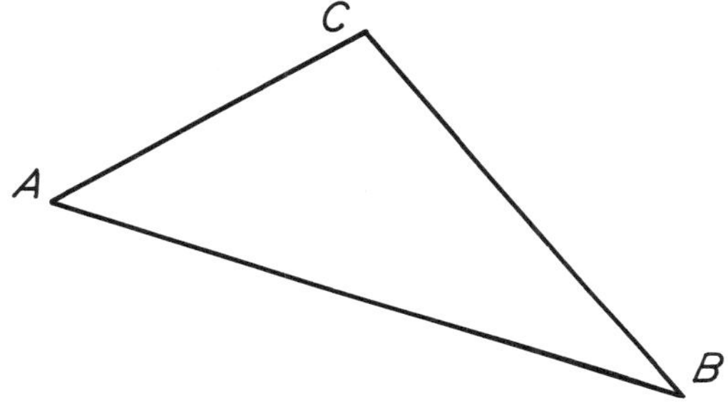


2 DETERMINE THE FRONT VIEW OF PLANE DEF. THE PLANE SLOPES 45° WITH A SLOPE DIRECTION OF S40°E. SHOW ALL CONSTRUCTION.

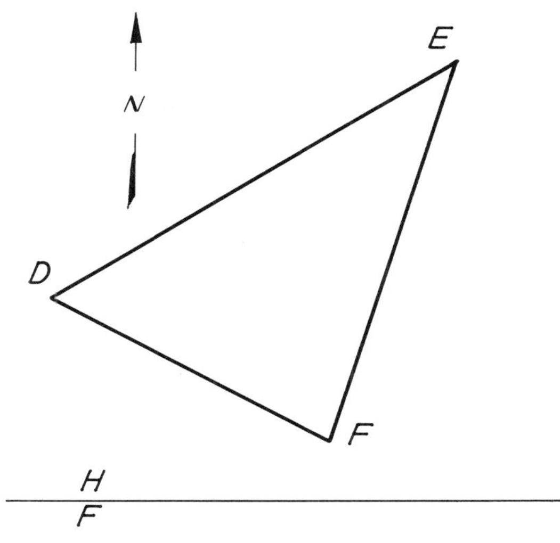


E

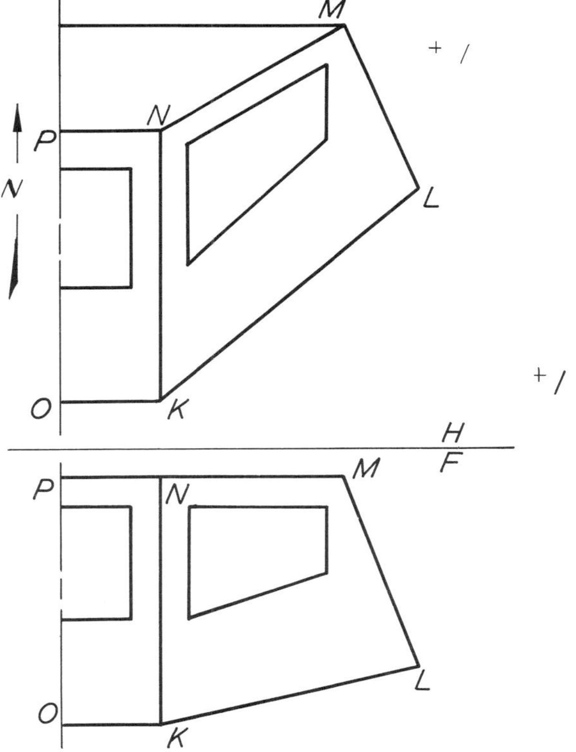


3 COMPLETE THE TABLE OF VALUES BELOW BY PROJECTING FROM THE VIEWS OF THE AIRCRAFT MOCK-UP.

| PLANE | SLOPE | SLOPE DIR |
|---|---|---|
| KLMN | | |
| OPNK | | |

Courtesy of Sikorsky Aircraft

# METRIC SCALES

**Remove this page and fold it longwise along the desired scale for making metric measurements.**

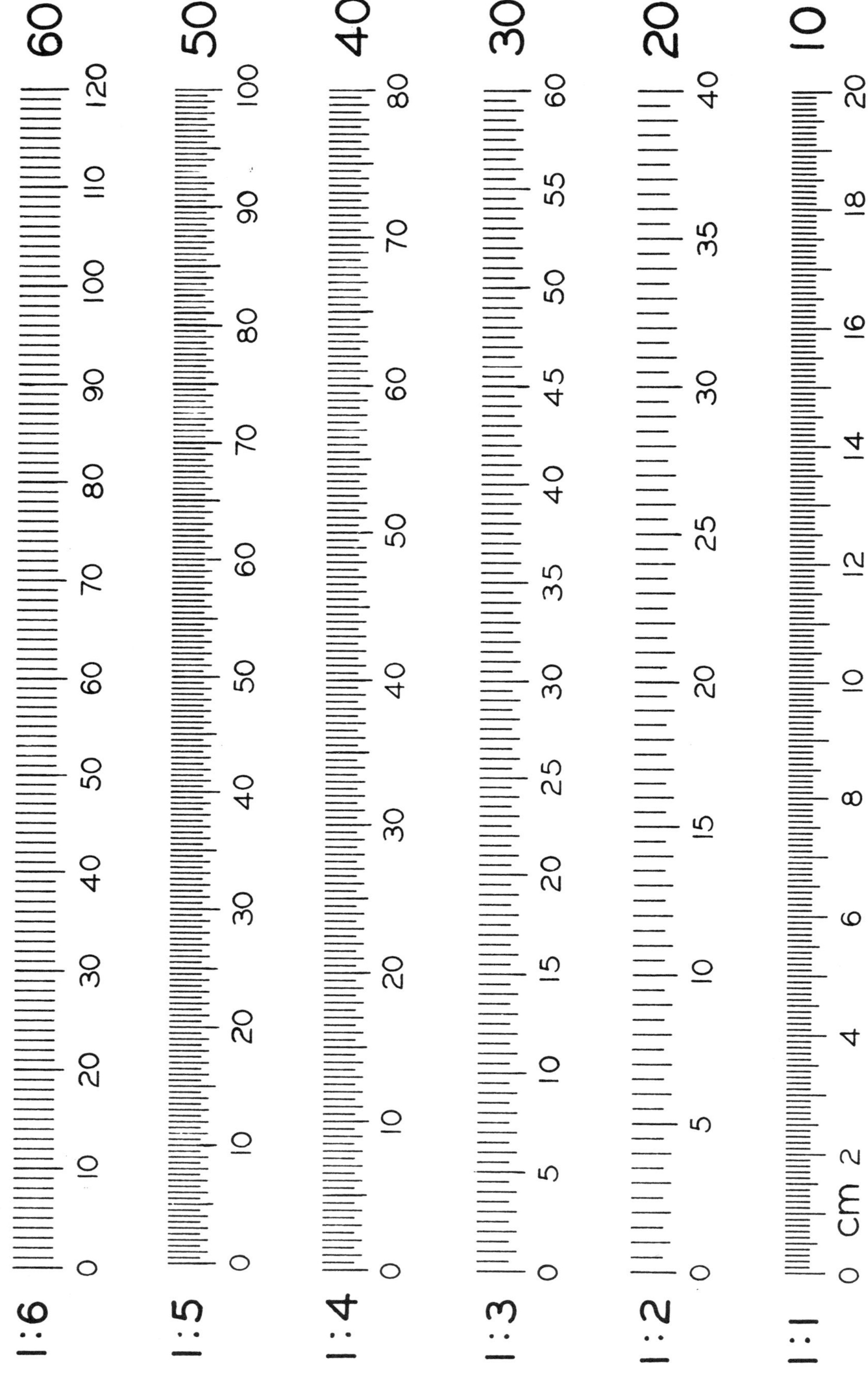

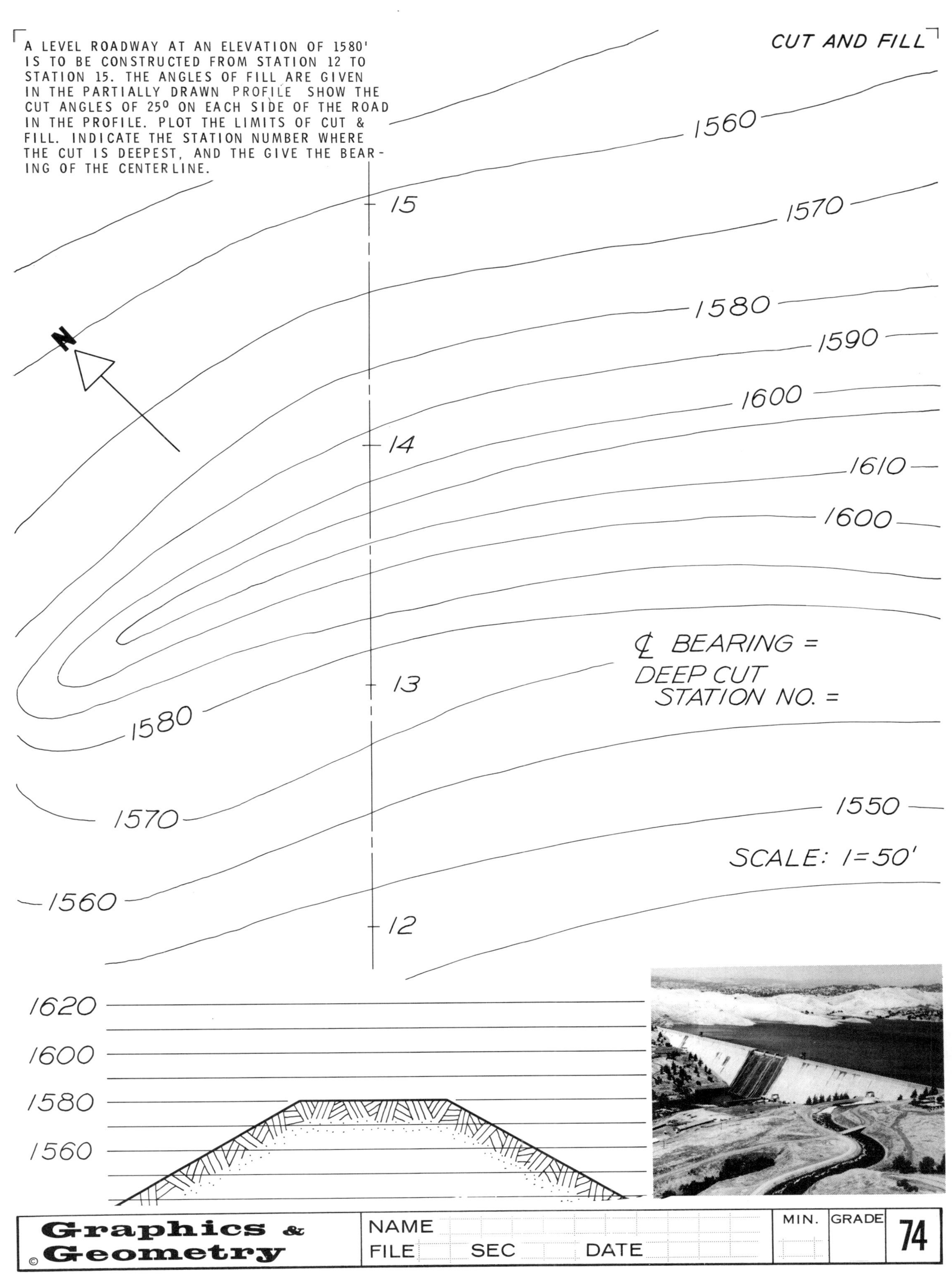

A LEVEL ROADWAY AT AN ELEVATION OF 1580' IS TO BE CONSTRUCTED FROM STATION 12 TO STATION 15. THE ANGLES OF FILL ARE GIVEN IN THE PARTIALLY DRAWN PROFILE SHOW THE CUT ANGLES OF 25° ON EACH SIDE OF THE ROAD IN THE PROFILE. PLOT THE LIMITS OF CUT & FILL. INDICATE THE STATION NUMBER WHERE THE CUT IS DEEPEST, AND THE GIVE THE BEARING OF THE CENTERLINE.
CUT AND FILL
1560
15
1570
1580
N
1590
1600
14
1610
1600
℄ BEARING =
DEEP CUT
STATION NO. =
13
1580
1570
1550
SCALE: 1=50'
1560
12
1620
1600
1580
1560
Graphics & Geometry
©
NAME
FILE
SEC
DATE
MIN.
GRADE
74

# METRIC SCALES

**Remove this page and fold it longwise along the desired scale for making metric measurements.**

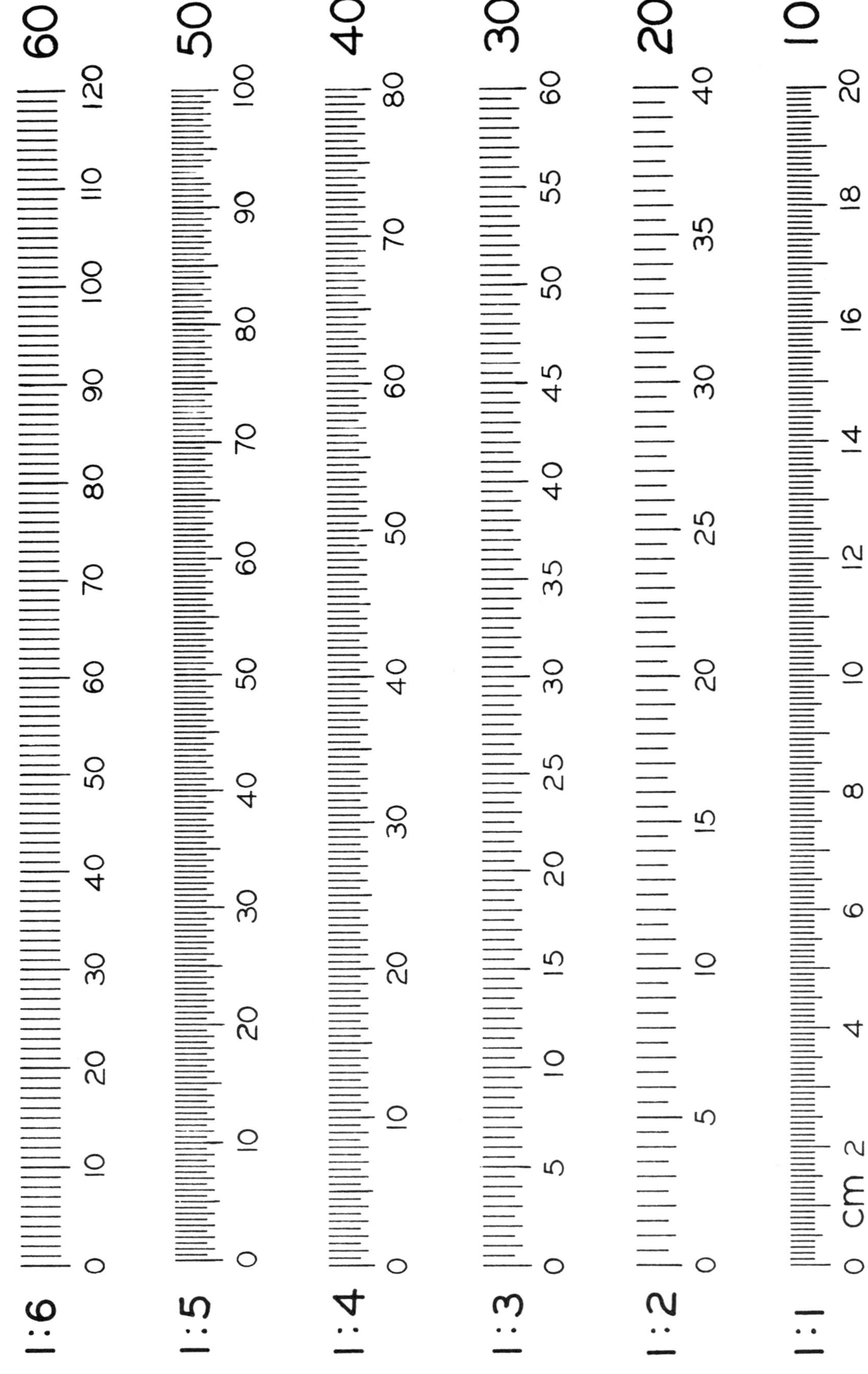

# EARTHEN DAM

A LEVEL EARTHEN DAM OF THE ELEVATION GIVEN IN THE PARTIALLY DRAWN PROFILE IS TO HAVE A CENTERLINE THAT FORMS AN ARC IN THE PLAN VIEW. COMPLETE THE PROFILE TO SHOW A SLOPE ANGLE OF 30° ON THE UPSTREAM SIDE AND THE SURFACE OF THE GROUND USING CUTTING PLANE A-A. SHOW THE LIMITS OF EARTHEN FILL IN THE PLAN VIEW AND THE LIMITS OF THE WATER AT AN ELEVATION OF 115 FEET.

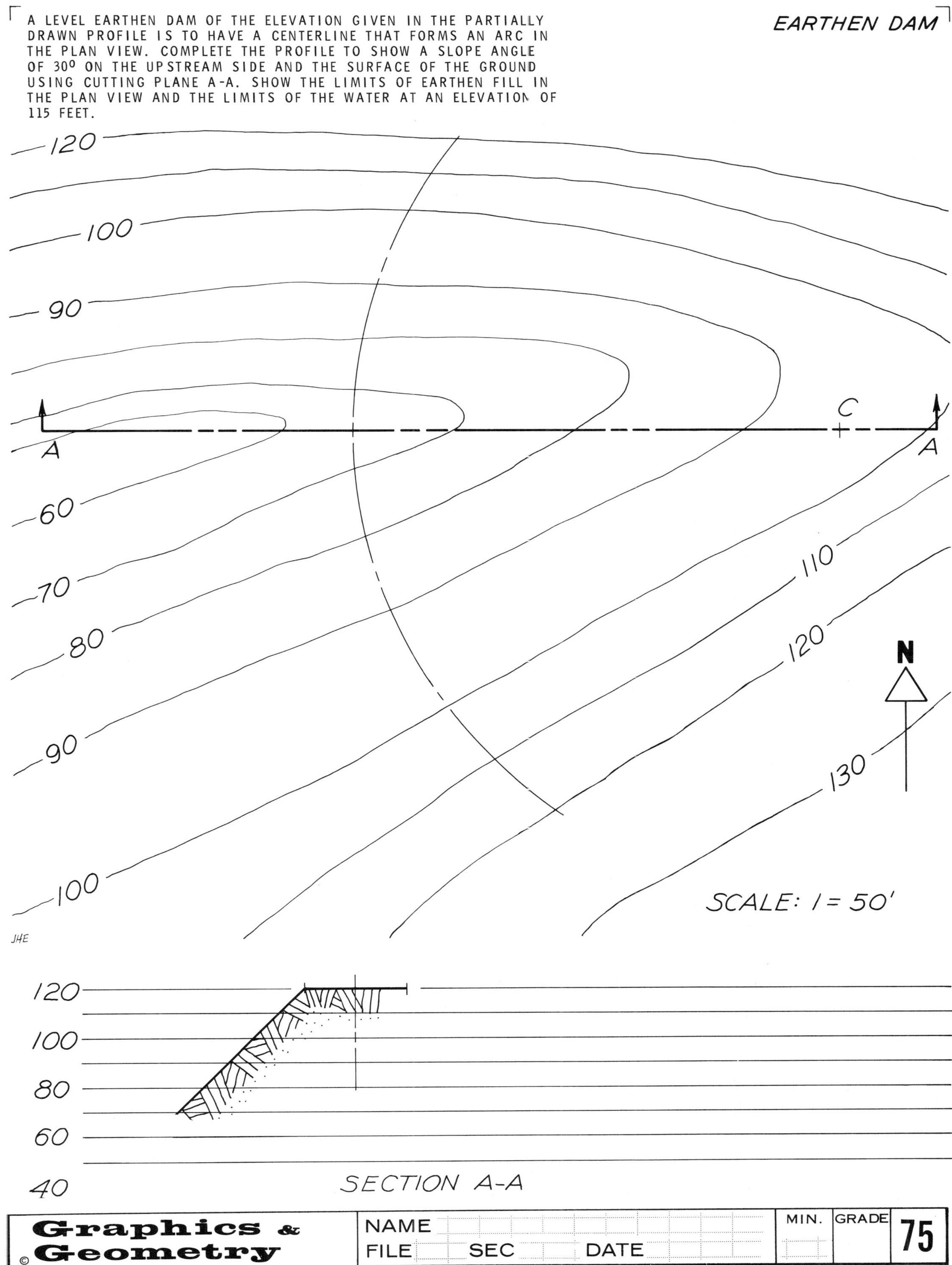

# METRIC SCALES

**Remove this page and fold it longwise along the desired scale for making metric measurements.**

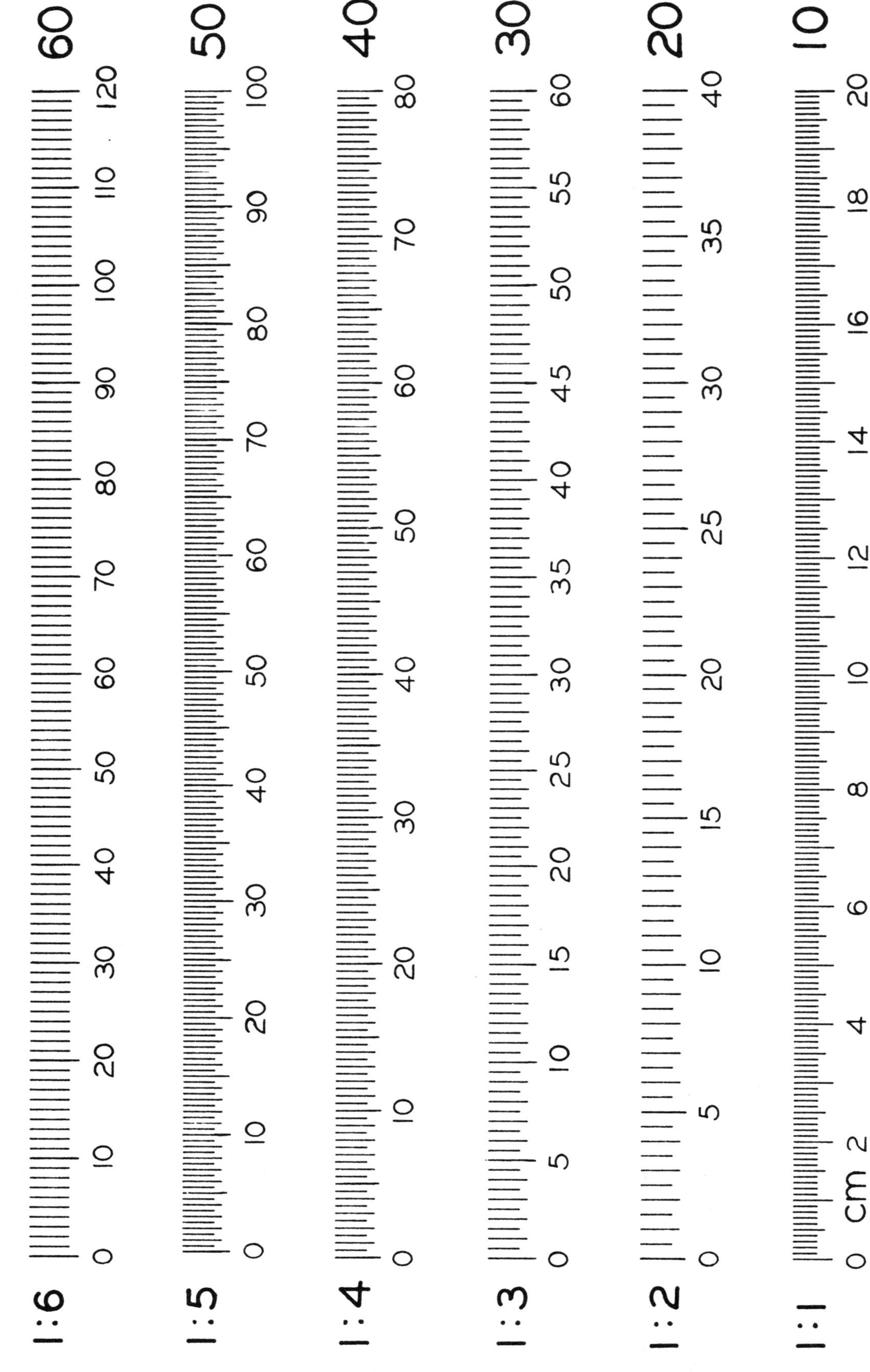

## STRIKE AND DIP OF A PLANE

1 DETERMINE THE STRIKE AND DIP OF PLANE ABC. SHOW CONSTRUCTION AND LABEL POINTS, ANGLES AND REFERENCE PLANES.

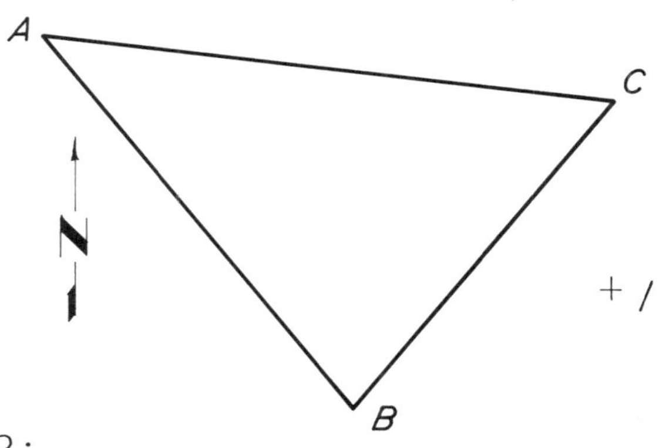


DIP:

STRIKE:

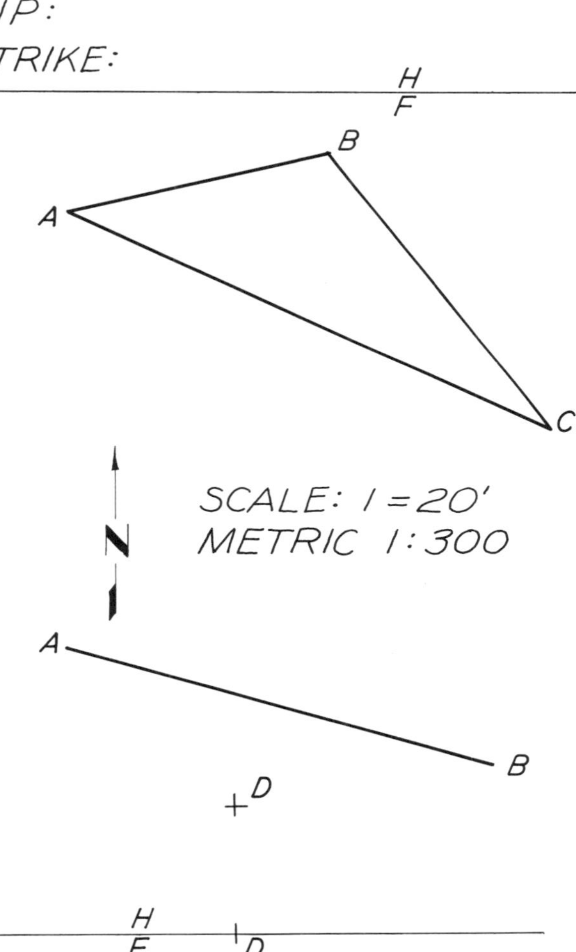

COURTESY ALLIS-CHALMERS

2 LINE AB IS THE STRIKE OF THE UPPER PLANE OF A MINERAL VEIN THAT DIPS 40°SW AND HAS A THICKNESS OF 25 FEET (10 METERS). CONSTRUCT AN AUXILIARY VIEW AND FIND THE VERTICAL DISTANCE FROM POINT D TO THE MINERAL VEIN. COMPLETE THE TABLE BELOW.

+ 1

A ——— B

STRIKE OF VEIN _

VERTICAL THICKNESS _

VERTICAL DISTANCE _

# METRIC SCALES

Remove this page and fold it longwise along the desired scale for making metric measurements.

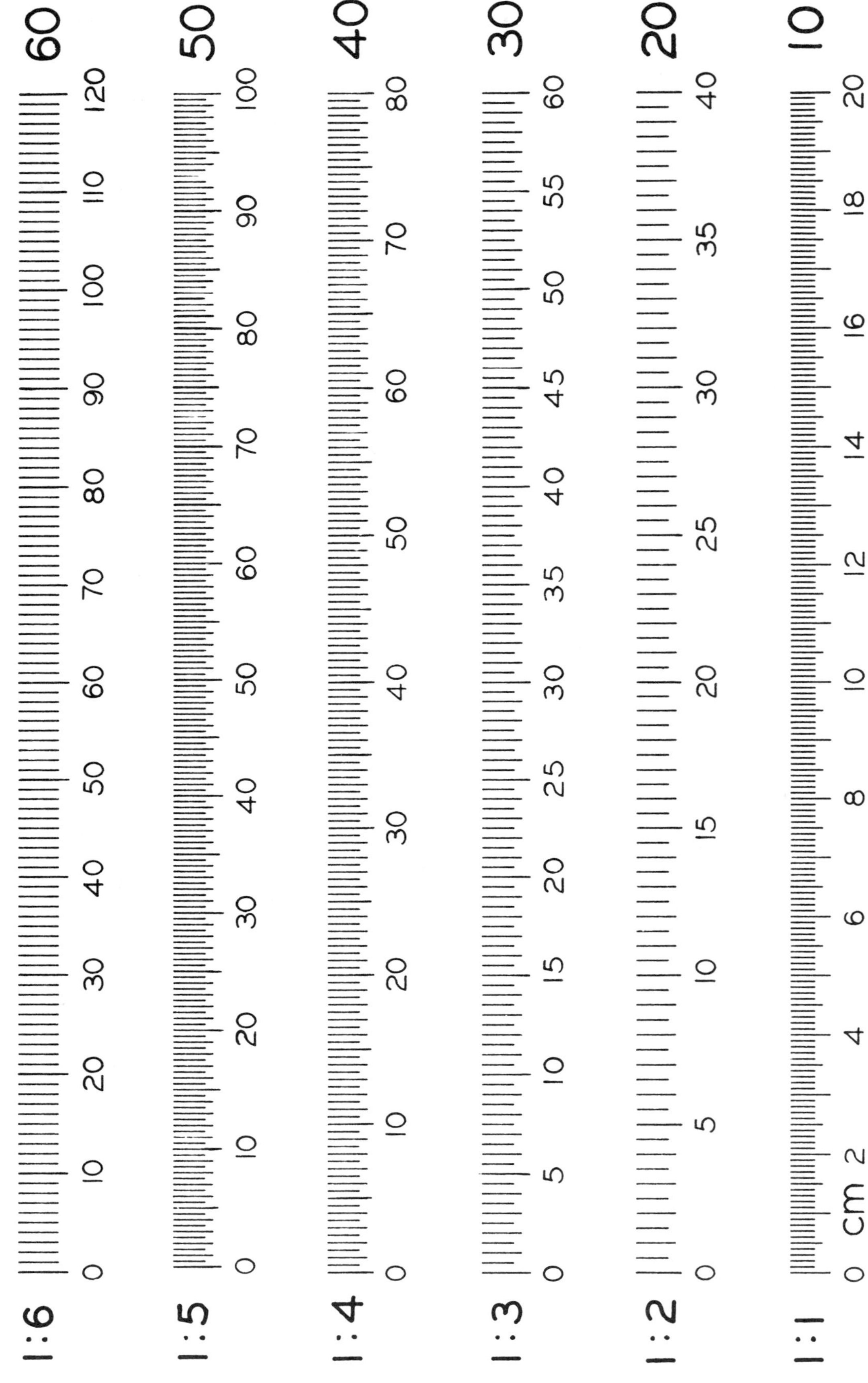

# OUTCROP

POINTS A, B, & C LIE ON THE SURFACE OF THE EARTH WHERE THE UPPER PLANE OF AN ORE VEIN OUTCROPS. A VERTICAL SHAFT AT D INTERSECTS THE BOTTOM PLANE AT AN ELEVATION OF 60 FT. COMPLETE THE PROFILE SECTION SHOWING THE SECTION THROUGH THE EARTH, THE ELEVATION LINES, THE ORE VEIN, THE SHORTEST AND VERTICAL SHAFTS FROM POINT E ON THE SURFACE OF THE EARTH. USE THE PROPER SYMBOLS IN THE PROFILE. COMPLETE THE TABLE OF VALUES.

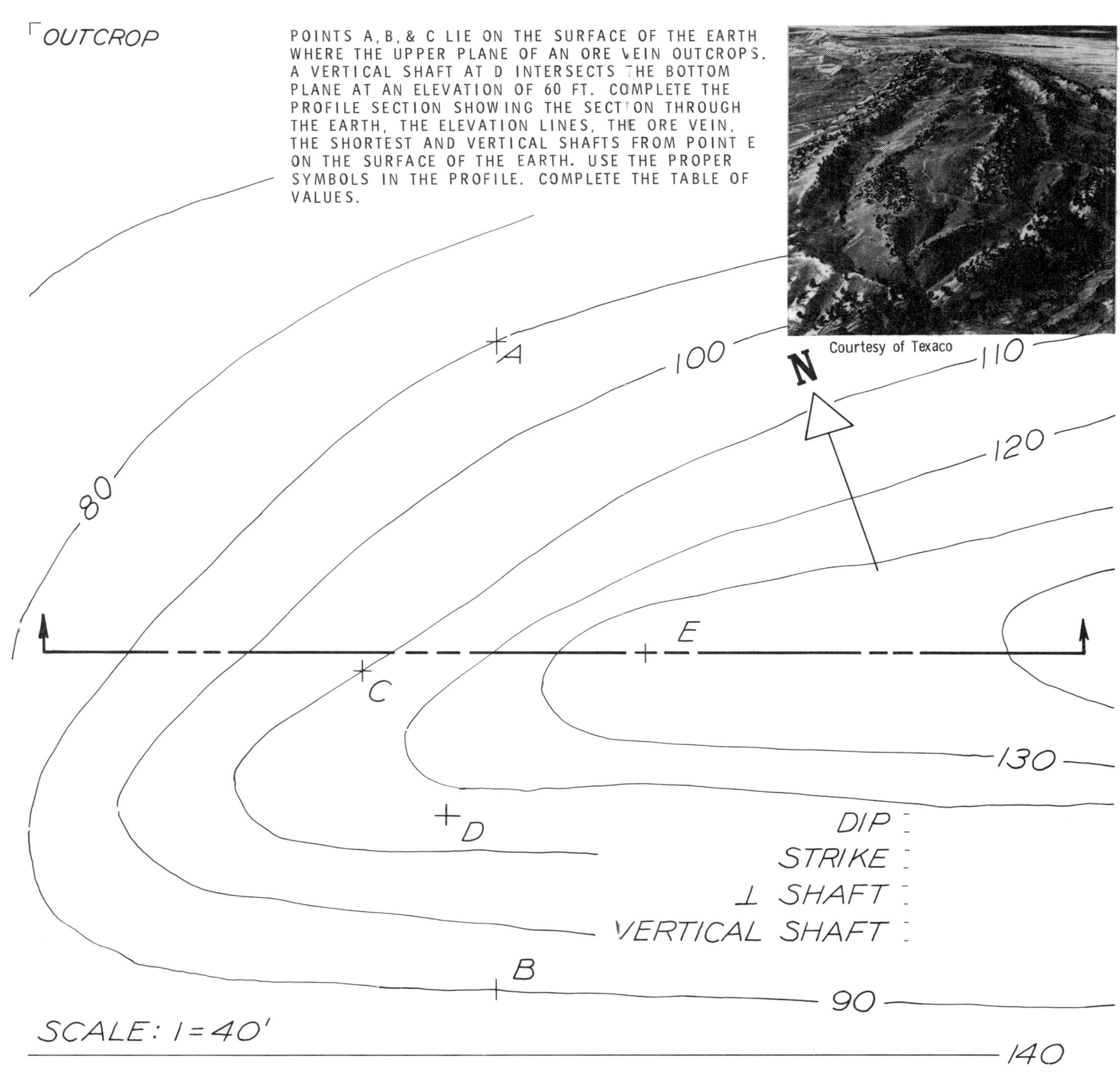

# METRIC SCALES

**Remove this page and fold it longwise along the desired scale for making metric measurements.**

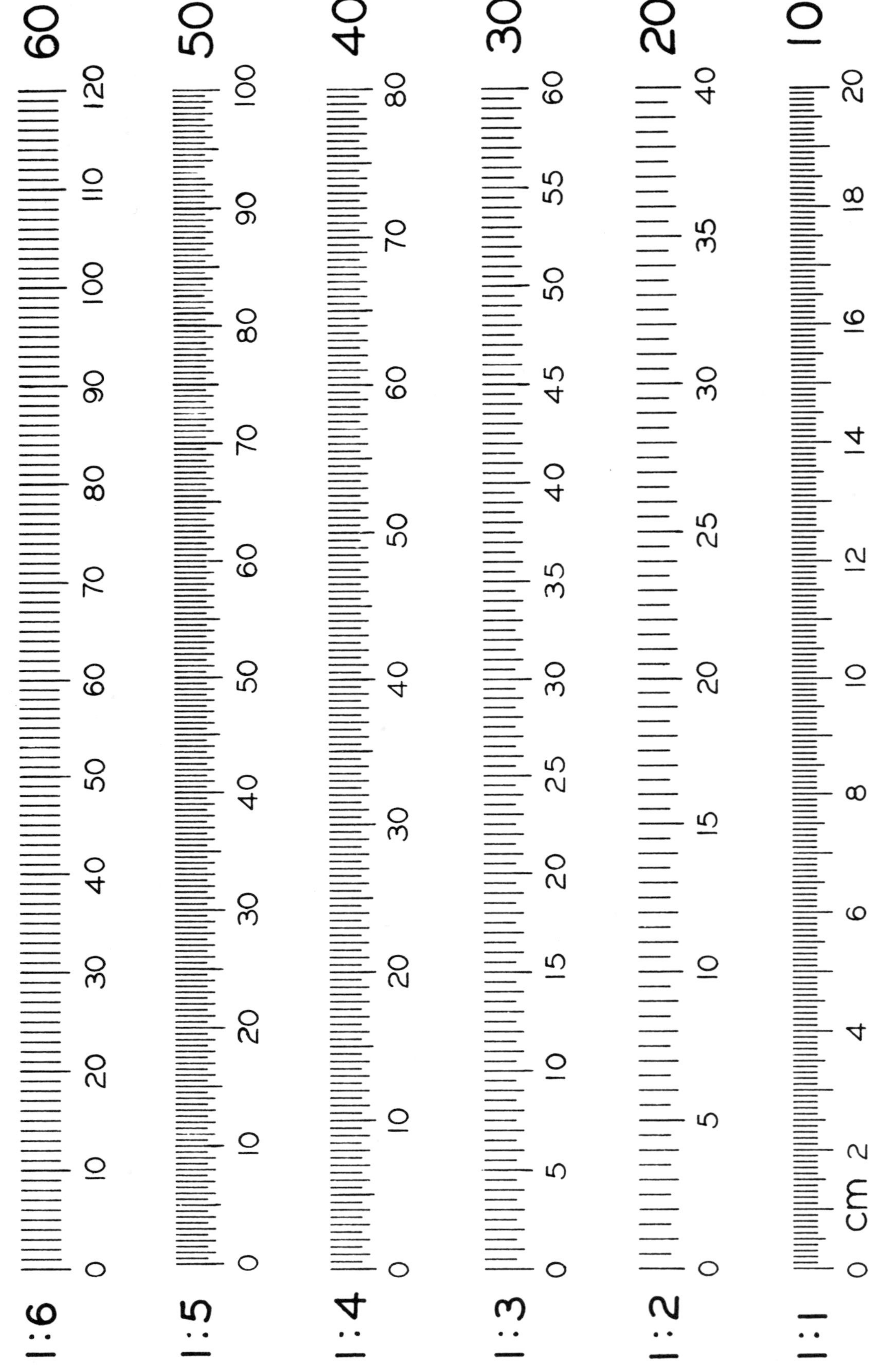

# INTERSECTION BETWEEN PLANES

**1** DETERMINE THE LINE OF INTERSECTION BETWEEN PLANES ABC AND XYZ, USING THE CUTTING PLANE METHOD. SHOW THE LINE OF INTERSECTION IN THE FRONT AND TOP VIEWS.

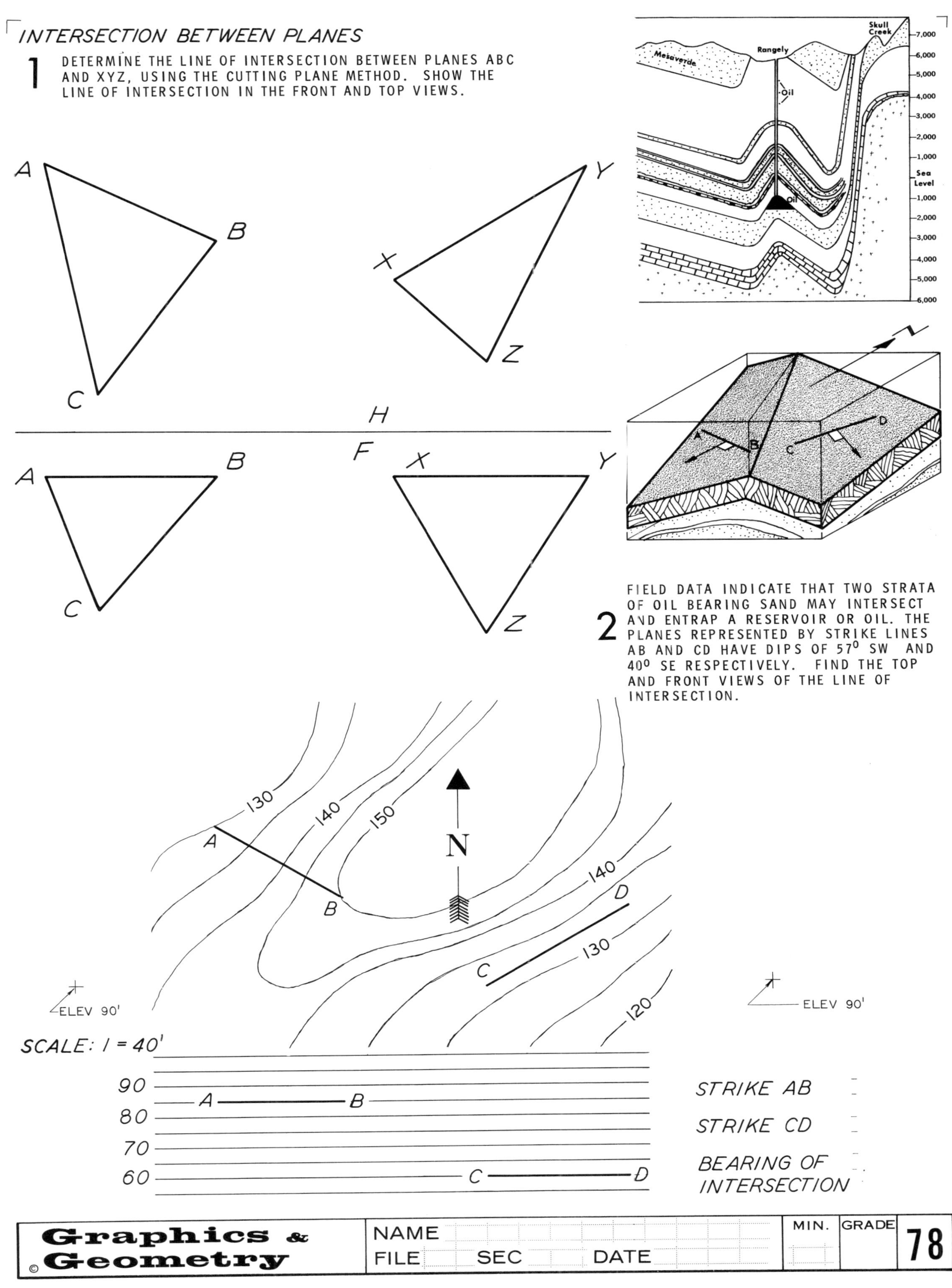


**2** FIELD DATA INDICATE THAT TWO STRATA OF OIL BEARING SAND MAY INTERSECT AND ENTRAP A RESERVOIR OR OIL. THE PLANES REPRESENTED BY STRIKE LINES AB AND CD HAVE DIPS OF 57° SW AND 40° SE RESPECTIVELY. FIND THE TOP AND FRONT VIEWS OF THE LINE OF INTERSECTION.

| Graphics & Geometry © | NAME | MIN. | GRADE | 78 |
|---|---|---|---|---|
| | FILE SEC DATE | | | |

# METRIC SCALES

Remove this page and fold it longwise along the desired scale for making metric measurements.

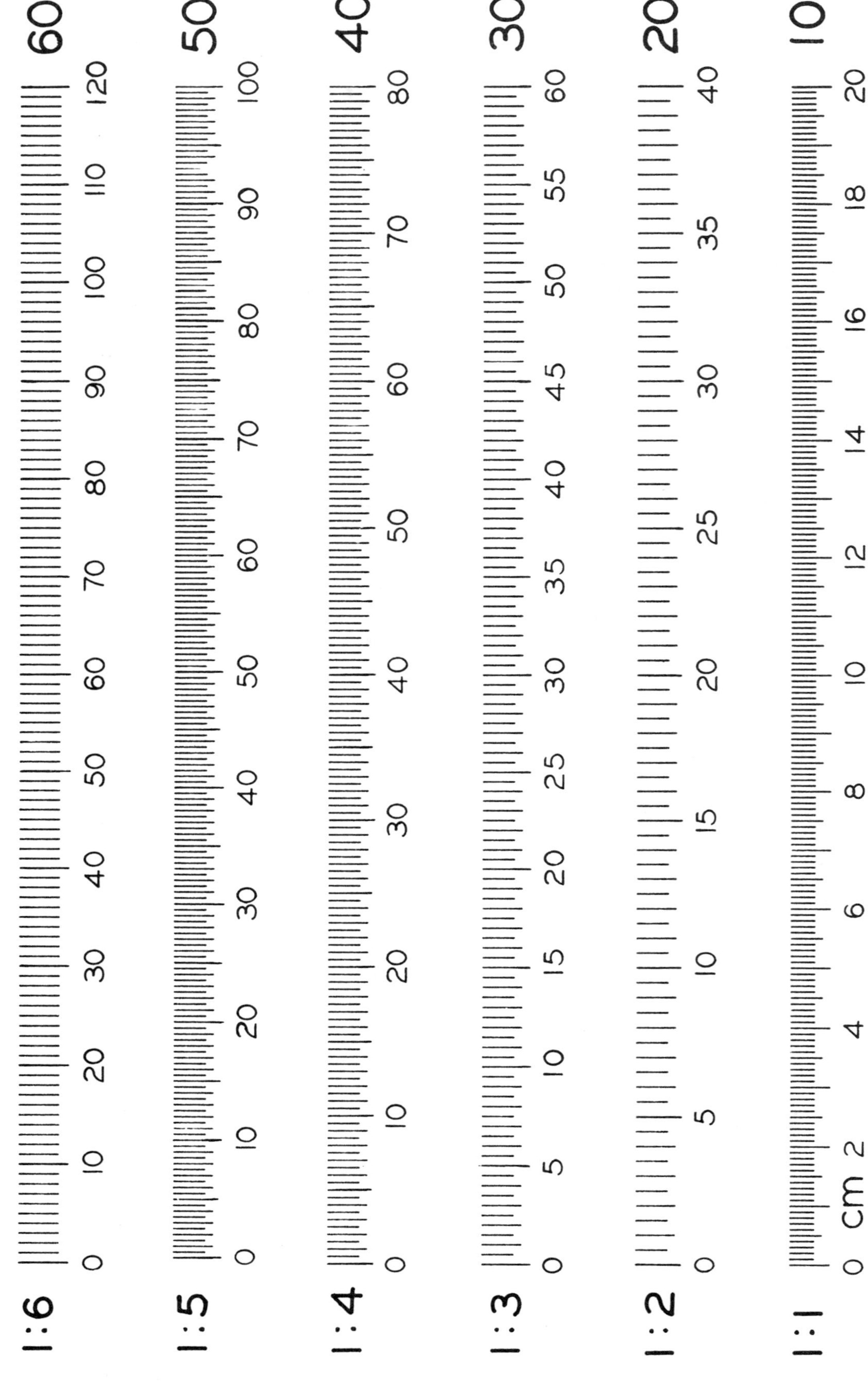

FIND THE POINT-VIEWS OF THE LINES BY AUXILIARY VIEWS AND LABEL ALL VIEWS.

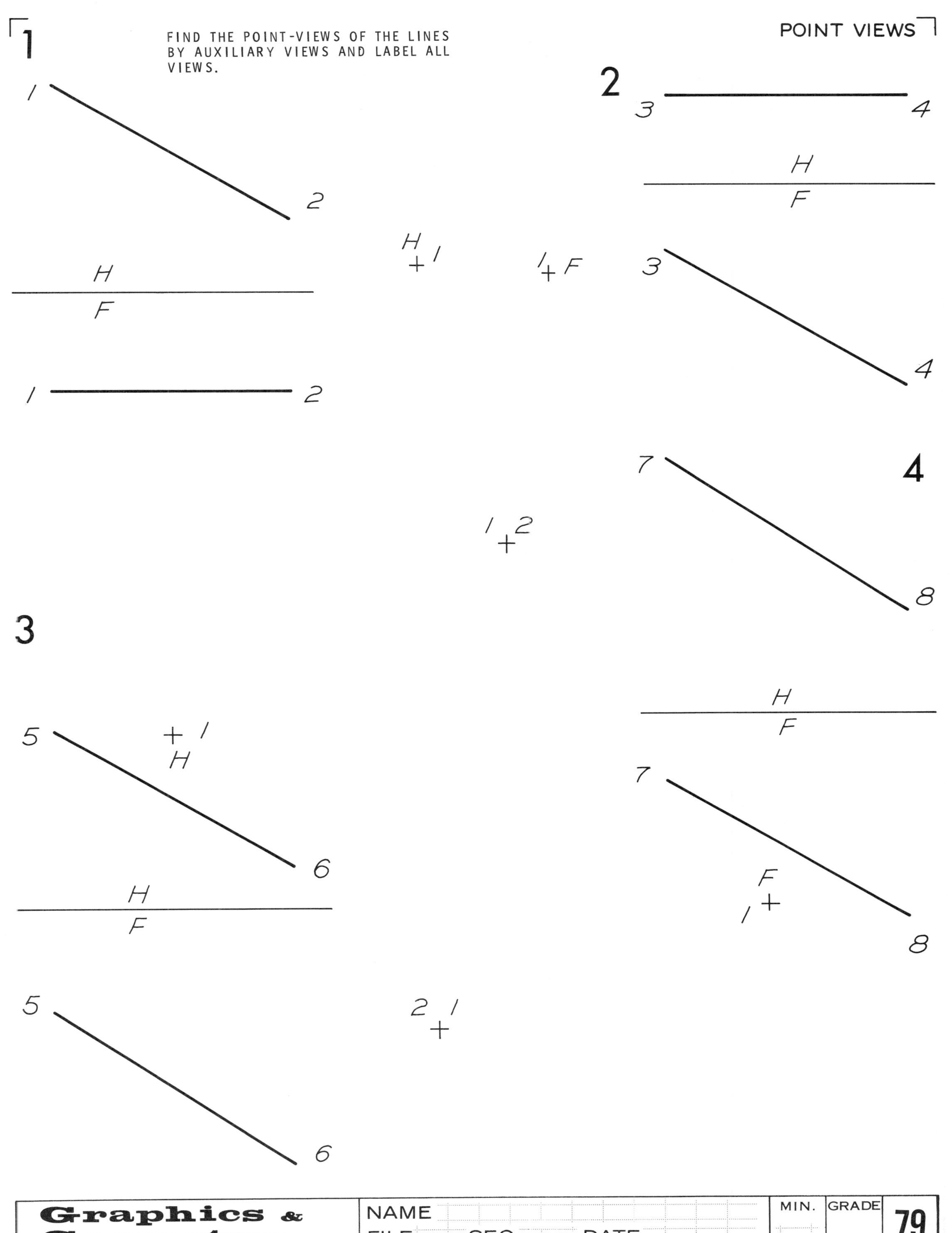

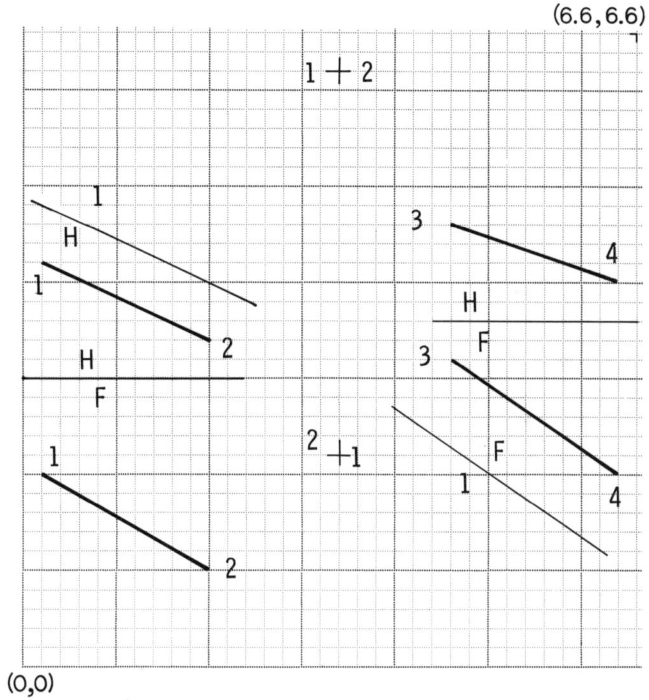

(6.6, 6.6)

1 + 2

1 H 1 2 H F 1 2

3 4 H F 3 F 1 4

2 + 1

(0,0)

Find the point views of the lines.

# ANGLE BETWEEN PLANES



**1** FIND THE ANGLES BETWEEN THE PLANES IN PROBLEMS 1 AND 2.

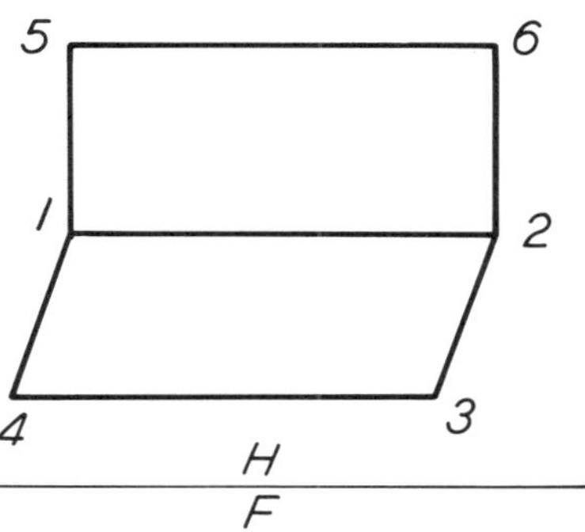


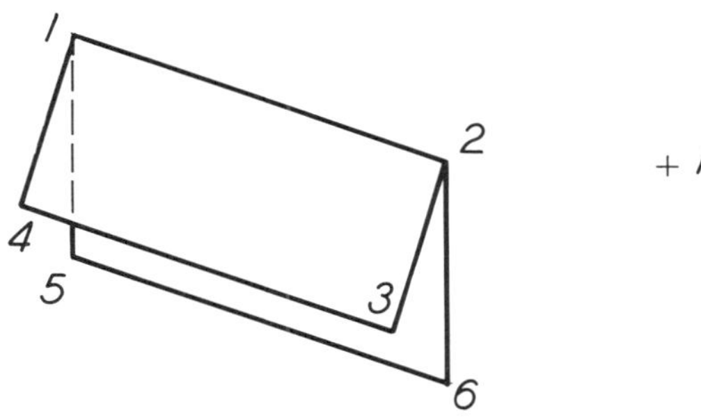


ANGLE =

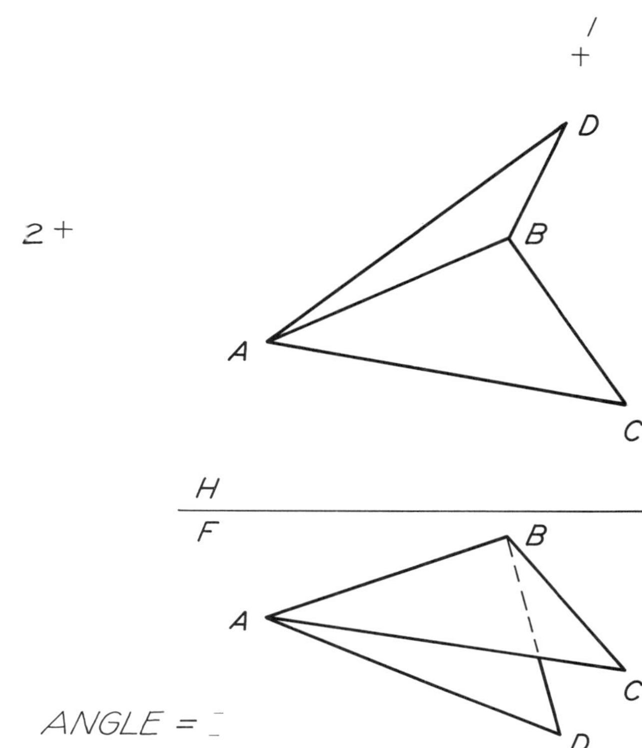


ANGLE =

**3** FIND THE ANGLE BETWEEN THE ROOF SURFACES THAT INTERSECT ALONG LINE 1-4. PROJECT FROM THE TOP VIEW.

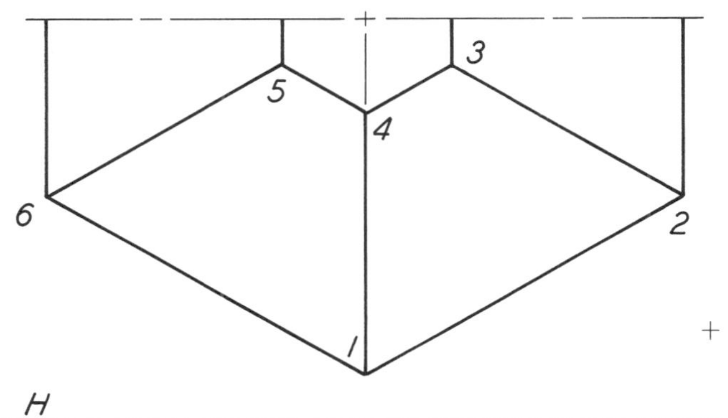

COURTESY AMERICAN IRON AND STEEL INSTITUTE

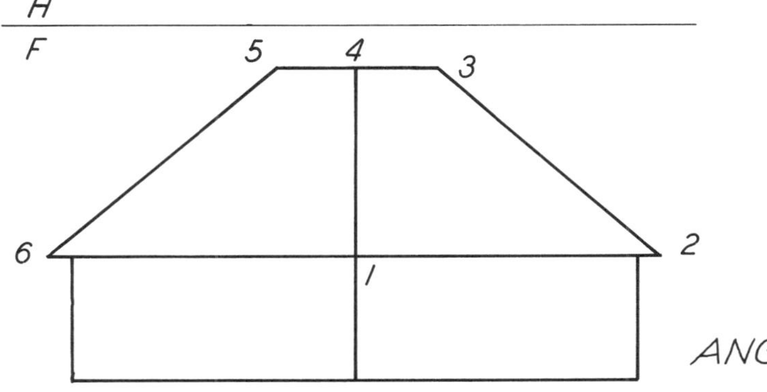


ANGLE:

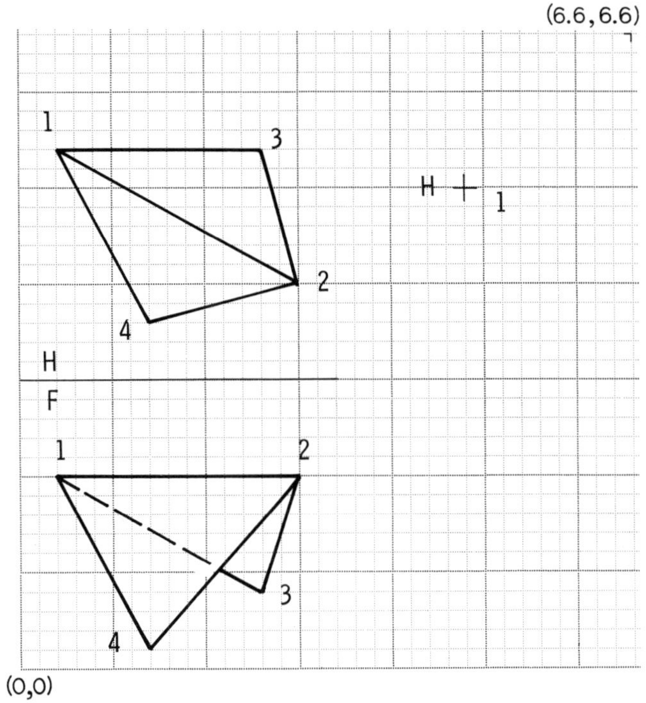

(6.6,6.6)

H | 1

H

F

(0,0)

Find the angle between the planes.

## TRUE SHAPE OF A PLANE

IN ORDER TO DESIGN A TRANSITION PIECE FOR A VENTILATING SYSTEM FIND THE FOLLOWING:

A. THE TRUE SHAPE OF PLANE 1-2-3-4

B. THE TRUE SHAPE OF PLANE 2-3-8-7

C. THE TRUE SHAPE OF PLANE 5-6-7-8

$2_A$ +

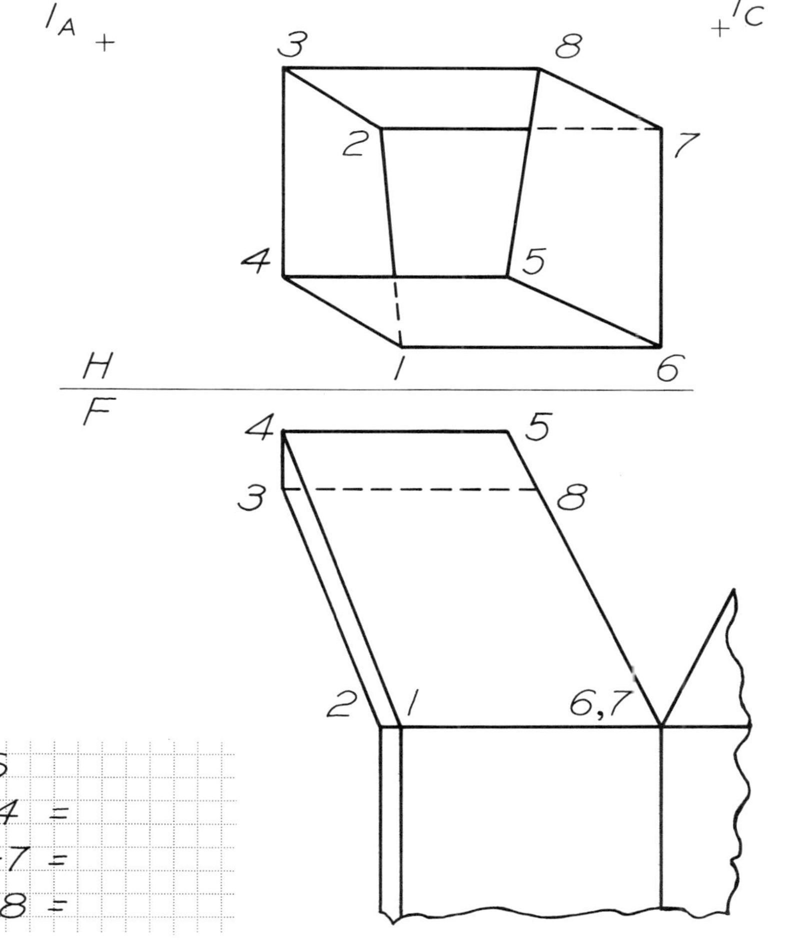


+ $2_C$

+ $1_B$

AREAS

1-2-3-4 =

2-3-8-7 =

5-6-7-8 =

SCALE: 1 = 2'

METRIC 1:30

| © Graphics & Geometry | NAME | | MIN. | GRADE | 81 |
|---|---|---|---|---|---|
| | FILE | SEC DATE | | | |

FIND THE TRUE SIZE VIEWS OF THE PLANES BELOW.

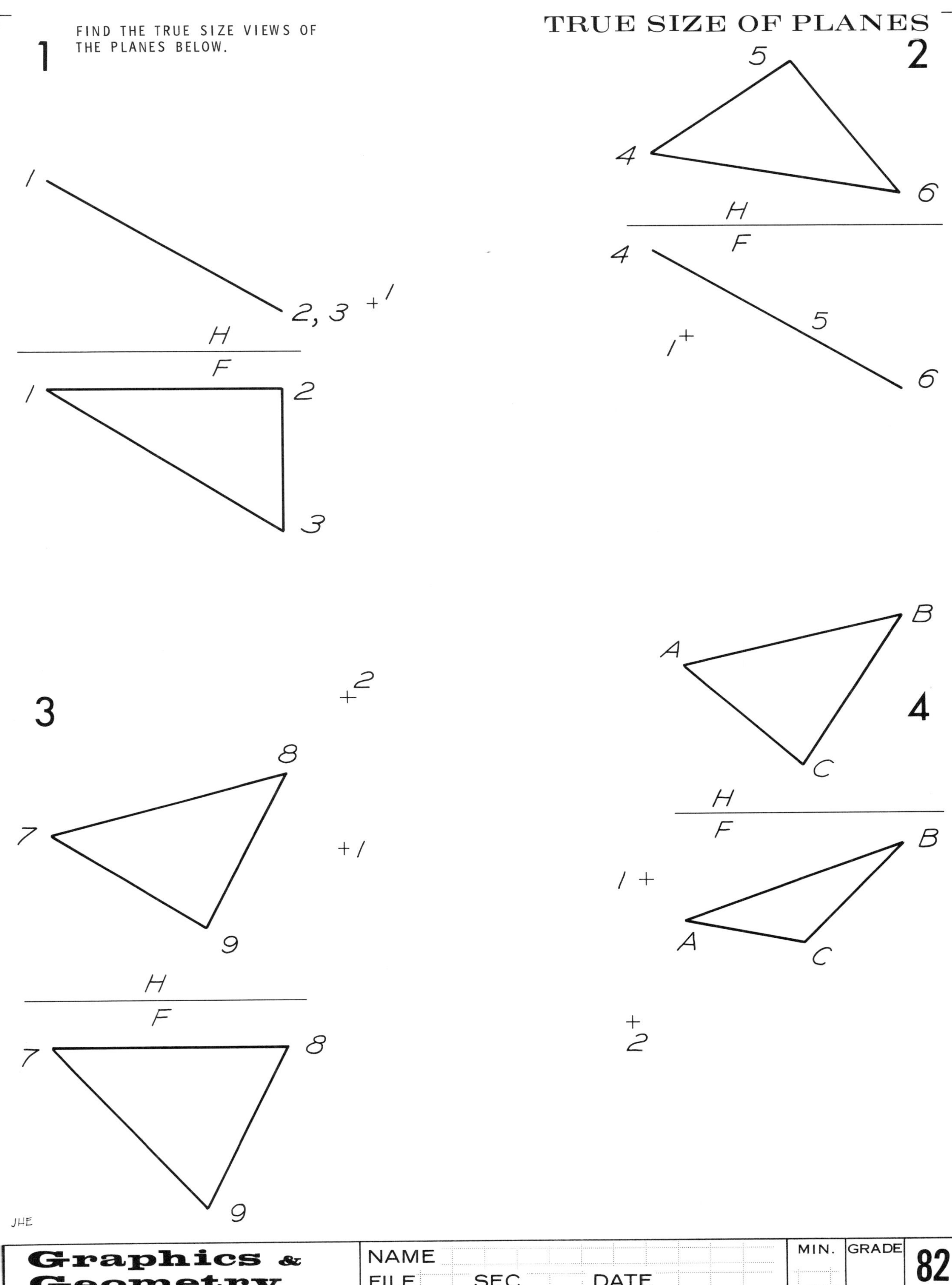


JHE

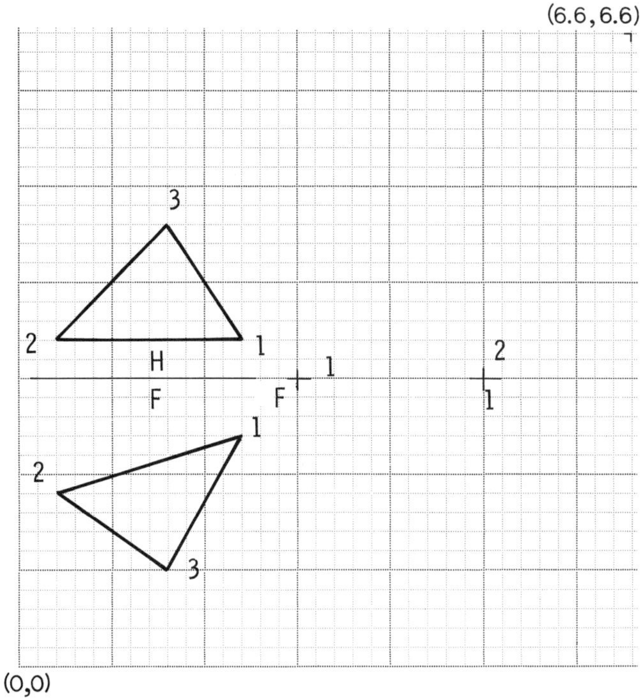

(6.6,6.6)

3
2 1
H
F F 1 2 1
1
2
3

(0,0)

Find the true-size view of the plane.

## ANGLE BETWEEN LINES

**1** FIND THE ANGLE BETWEEN LINES 1-2 AND 2-3 BY AUXILIARY VIEWS.

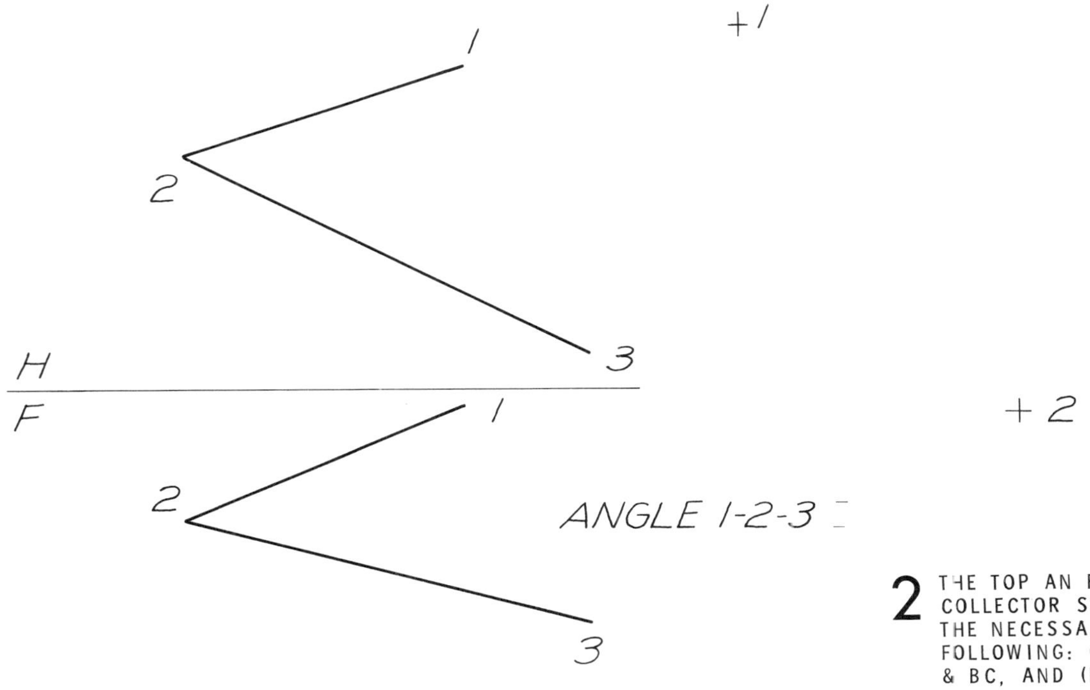

+1

1

2

3

H

F

1

2

3

+2

ANGLE 1-2-3 =

**2** THE TOP AN FRONT VIEWS OF A DUST COLLECTOR SYSTEM ARE GIVEN. DRAW THE NECESSARY VIEWS TO FIND THE FOLLOWING: (A) ANGLES BETWEEN AB & BC, AND (B) ANGLE BETWEEN CB & BD.

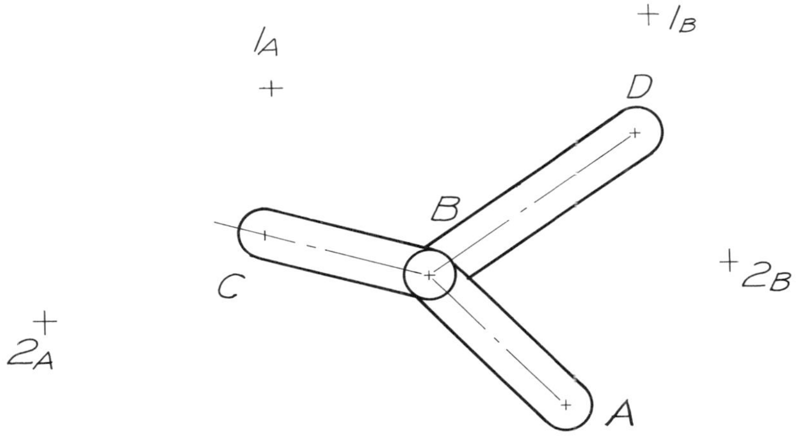

+ $1_B$

$1_A$ +

D

B

C

+ $2_B$

+ $2_A$

A

A. ANGLE ABC:

B. ANGLE CBD:

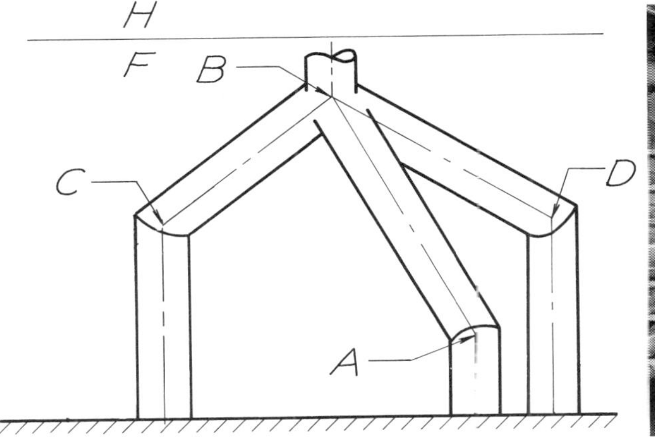

## SHORTEST DISTANCE – POINT TO A LINE

SCALE: 1 = 20′
METRIC: 1:30

2

**1** USING THE LINE METHOD, FIND THE SHORTEST DISTANCE FROM POINT O TO THE TWO LINES, AND SHOW THE LINES IN ALL VIEWS.

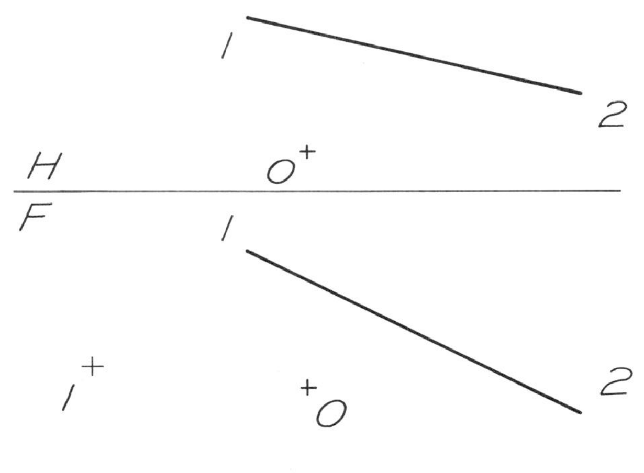


+2

DISTANCE:

1
2
O
3
4
H
F
4
3
O

DIST =

**3** THE ELECTRICAL CONDUIT, AB, IS SUPPORTED ON THE UPPER PART OF THE NUMERICALLY CONTROLLED MACHINE BY A SWIVEL BRACKET AT P. FIND THE SHORTEST DISTANCE FROM P TO THE CONDUIT BY THE LINE METHOD AND SHOW IT IN ALL VIEWS.

P
B
A
1
2

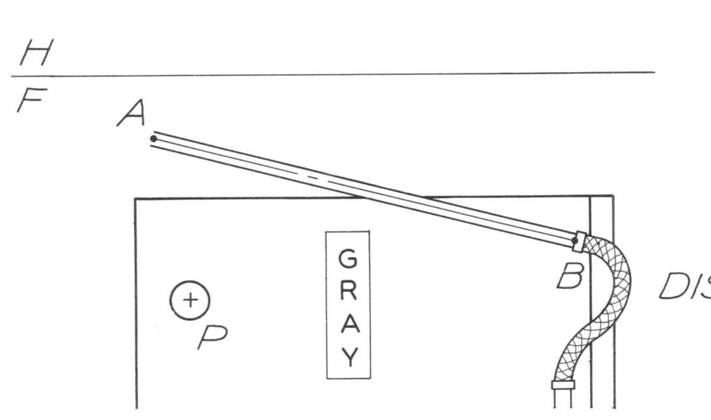


DIST =

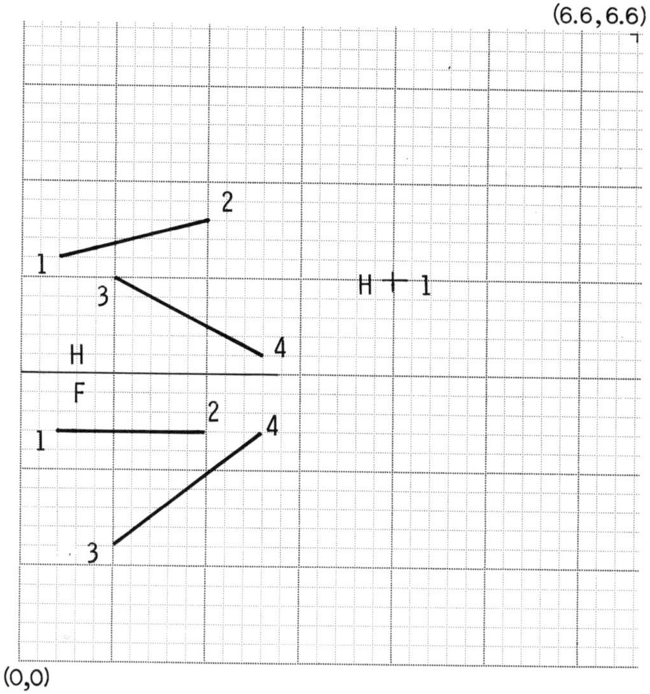
(6.6,6.6)
2
1
3
H + 1
4
H
F
2
4
1
3
(0,0)

Find the shortest distance between the two lines.

| | | | | |
|---|---|---|---|---|
| COMPUTER GRAPHICS | **Graphics for Engineers** © | NAME<br>FILE SEC DATE | MIN | GRADE | 84 |

# SKEWED LINES—LINE METHOD

FIND THE SHORTEST CONNECTING LINE BETWEEN AB AND 1-2. SHOW IN ALL VIEWS. **1**

**2** USING LINES DE AND 3-4 AS LINES ON ANGLE-IRON BRACES, FIND THE SHORTEST DISTANCE BETWEEN THEM BY THE LINE METHOD. FIND LINE DE TRUE LENGTH AS THE FIRST STEP. SHOW THE SHORETEST CONNECTOR IN ALL VIEWS.

A
2
1
B
1
H
F
B
A
1 2

SCALE: 1 = 20′
METRIC 1:300
ANSWER:

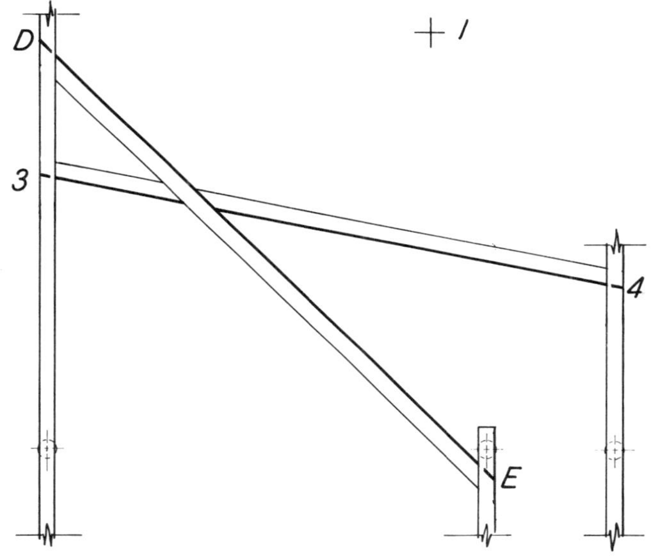


2

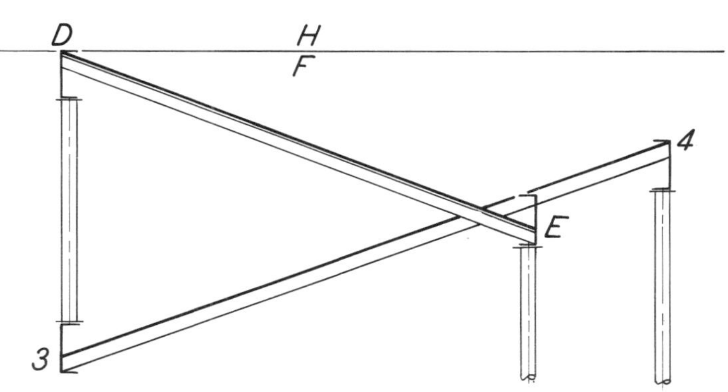


SCALE: 1 = 2′
METRIC 1:30
ANSWER:

## *SKEWED LINES—PLANE METHOD*

TWO SEGMENTS OF HIGHWAYS ARE GIVEN. FIND THE FOLLOWING:

1. THE SHORTEST HORIZONTAL (LEVEL) DISTANCE BETWEEN THEIR CENTER LINES AND SHOW IN ALL VIEWS.
2. FIND THE BEARING OF THIS LEVEL CONNECTOR?
3. FIND THE SHORTEST VERTICAL CLEARANCE BETWEEN THE HIGHWAYS? SHOW IN ITS TRUE-LENGTH VIEW.

SCALE: 1=20'
METRIC 1:300

HORIZONTAL LINE
LENGTH _
BEARING _
VERT. DIST. _

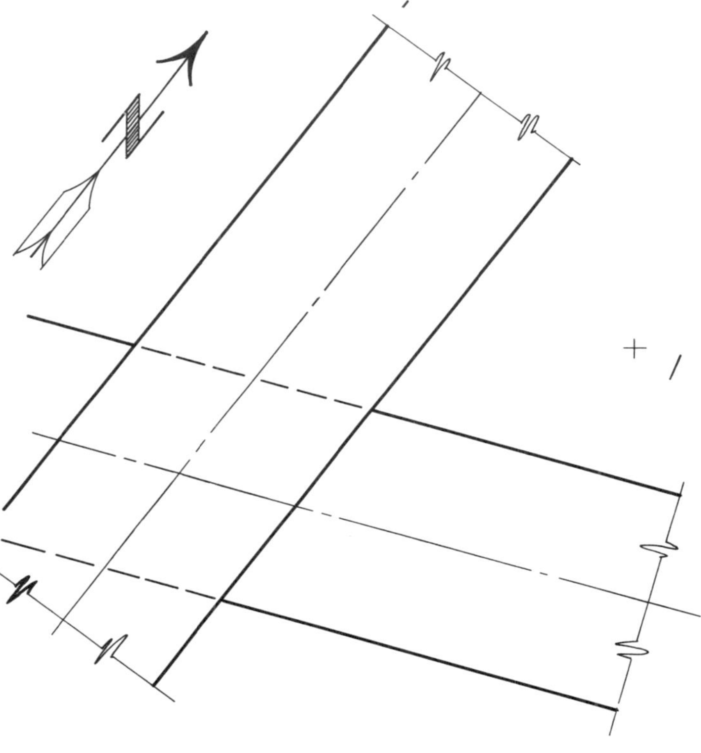



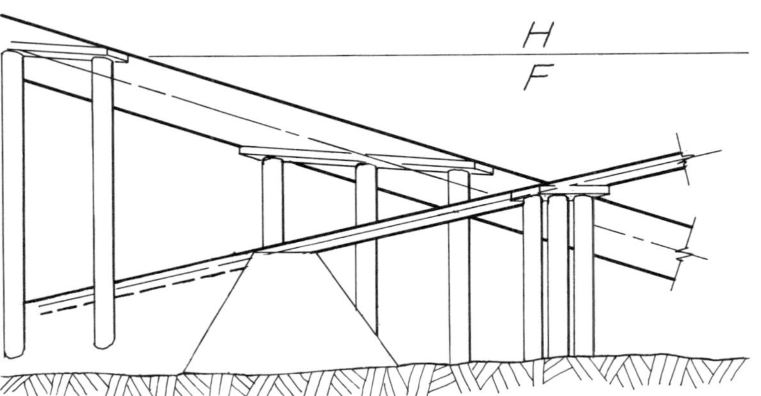

# ANGULAR DISTANCE–POINT TO A LINE

SCALE: 1=2'
METRIC 1:25

1 BY THE PLAINE METHOD, FIND THE SHORTEST PIPE THAT SLOPES UPWARD FROM POINT O TO CONNECT WITH PIPE 1-2. SHOW THE CONNECTION IN ALL VIEWS.

2 A PORTION OF A CATWALK FROM A TOWER TO A DRILLING PLATFORM IS GIVEN. FIND THE ANGLE BETWEEN BRACE 1-2 AND THE UPPER STRUT.

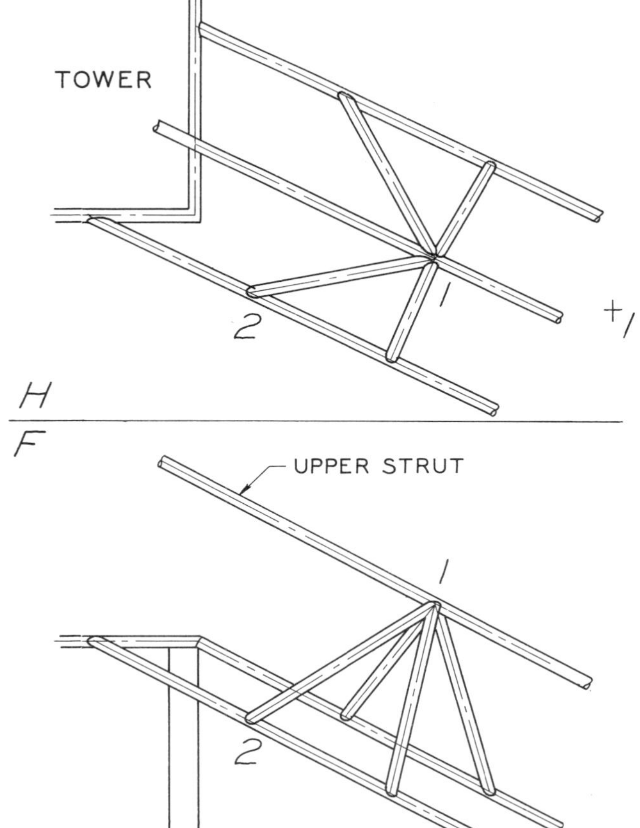


SCALE: 1=20'
METRIC: 1:300

LENGTH 1-2:
ANGLE:

| Graphics & Geometry © | NAME | | MIN. | GRADE | 87 |
|---|---|---|---|---|---|
| | FILE | SEC DATE | | | |

# ANGLE BETWEEN A LINE AND A PLANE

**1** FIND THE ANGLE BETWEEN THE LINE AND PLANE BY THE PLANE METHOD. SHOW THE VISIBILITY IN ALL VIEWS.

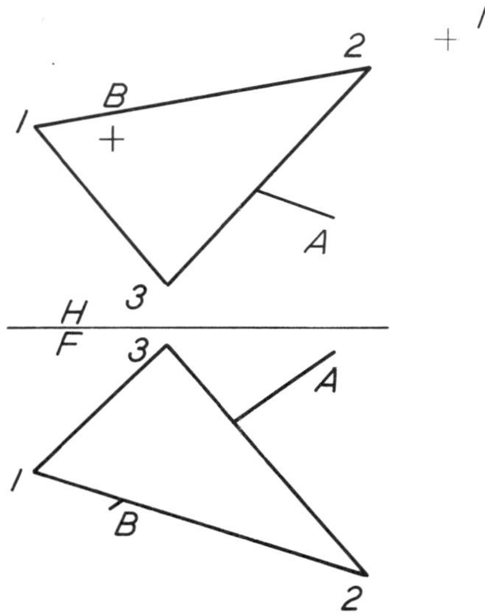


1

2

ANGLE:

**2**

FIND THE ANGLE OF INTERSECTION BETWEEN BRACE C AND THE PLANE DEFINED BY BRACES A AND B. USE THE PLANE METHOD.

3

3

SCALE: 1=10

OR 1:10

A

B

C

O

H

F

O

B

A

C

2

1

ANGLE:

| Graphics & Geometry © | NAME | | MIN. | GRADE | 88 |
|---|---|---|---|---|---|
| | FILE | SEC DATE | | | |

## REVOLUTION

**1** FIND THE TRUE LENGTH OF 1-2 IN THE FRONT VIEW BY REVOLVING THE LINE ABOUT AN AXIS THROUGH POINT 1.

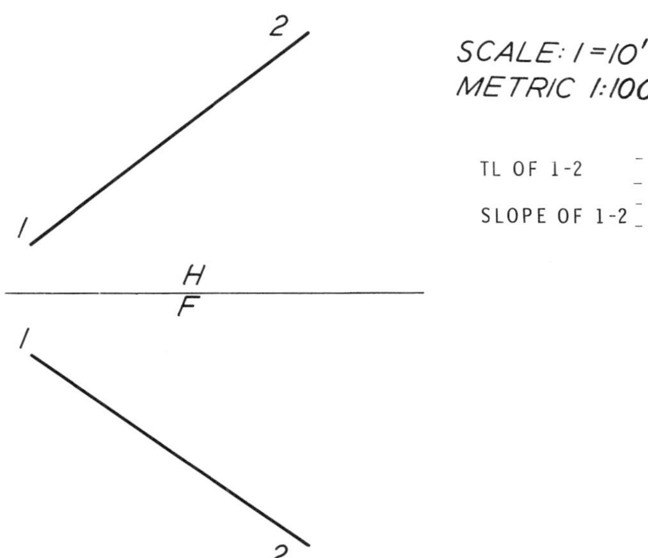


SCALE: 1 = 10'
METRIC 1:100

TL OF 1-2 ___

SLOPE OF 1-2 ___

**2** FIND THE TRUE LENGTH OF AB IN THE TOP VIEW BY REVOLVING THE LINE ABOUT AN AXIS THROUGH POINT A. FIND THE ANGLE WITH THE FRONTAL PLANE.

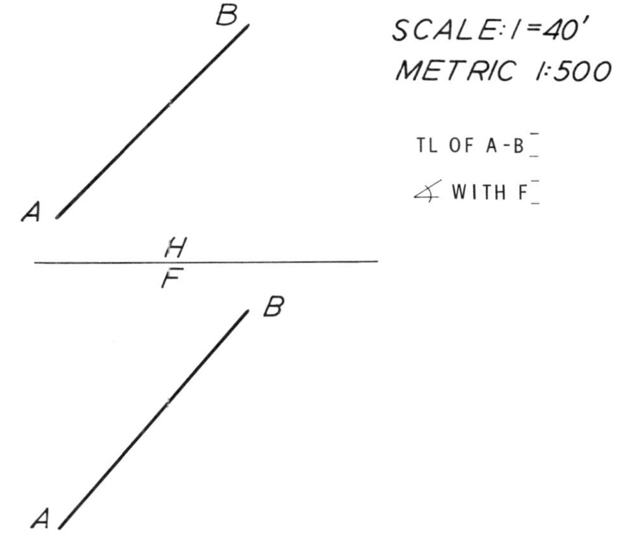


SCALE: 1 = 40'
METRIC 1:500

TL OF A-B ___

∡ WITH F ___

**3** FIND THE EDGE VIEW OF THE PLANE BY AN AUXILIARY VIEW, THEN FIND THE PLANE TRUE SIZE BY REVOLUTION.

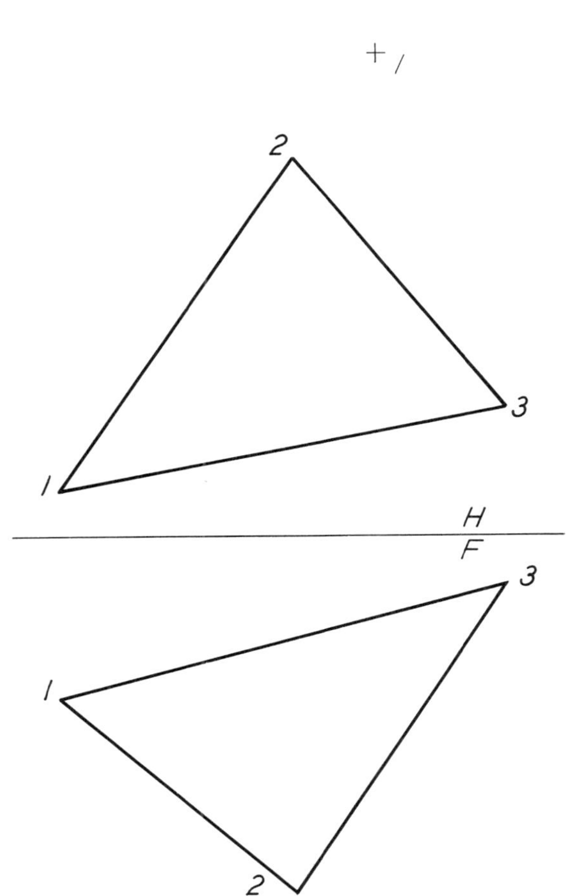


**4** FIND THE TRUE SIZE VIEW OF THE PLANE IN THE TOP VIEW BY DOUBLE REVOLUTION.

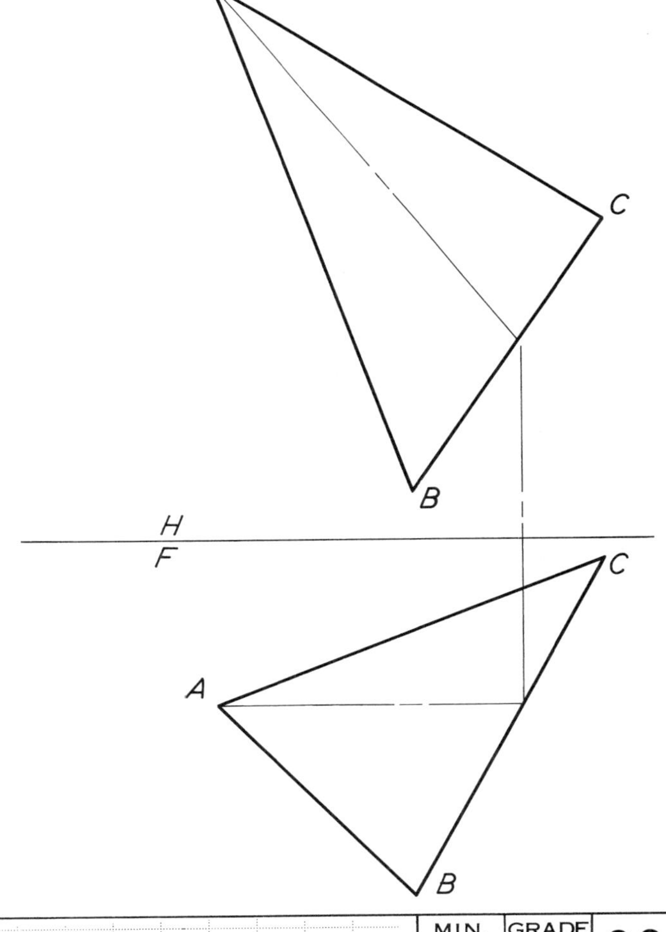

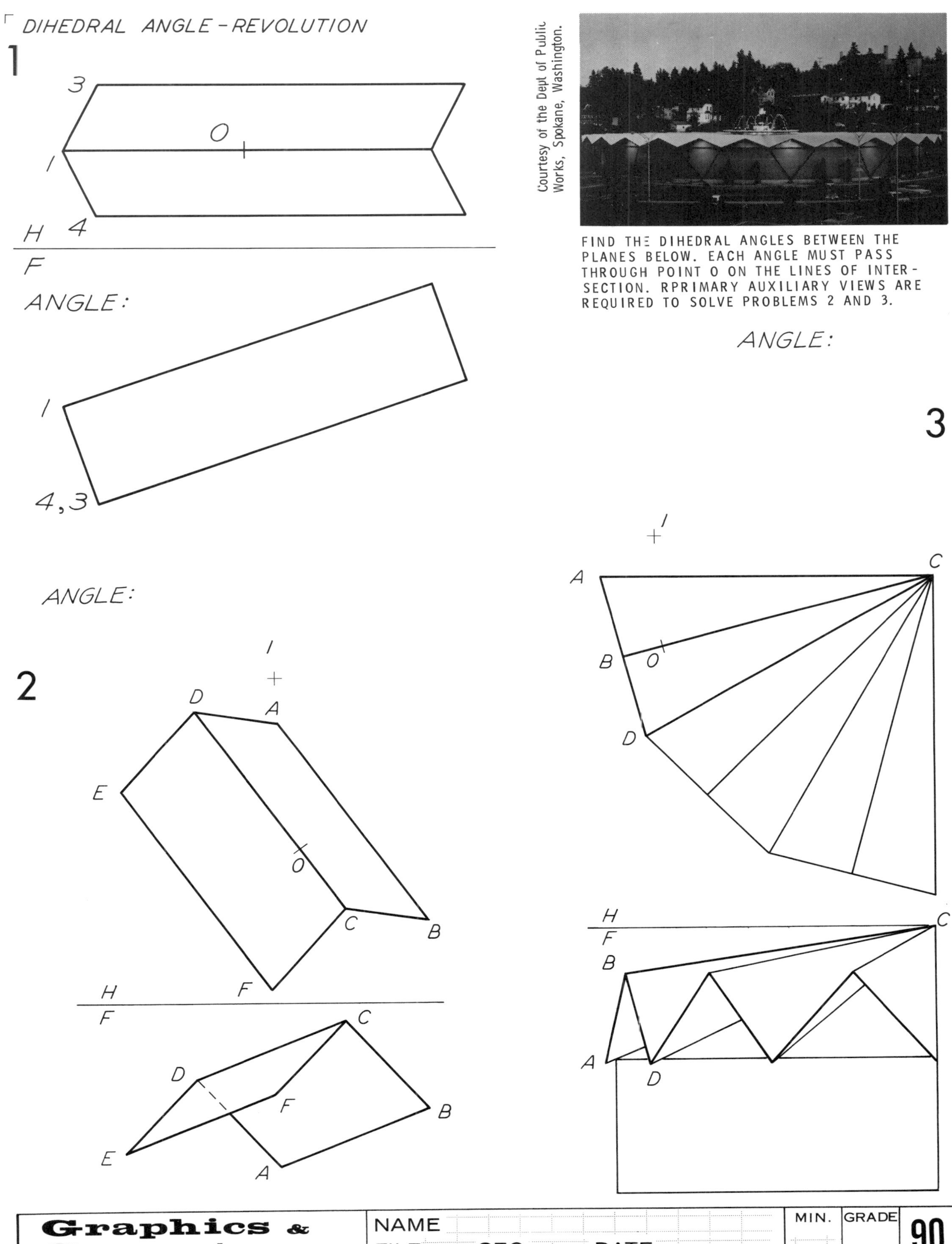
DIHEDRAL ANGLE - REVOLUTION
1
3
O
1
H 4
F
ANGLE:
1
4,3
ANGLE:
Courtesy of the Dept of Public Works, Spokane, Washington.
FIND THE DIHEDRAL ANGLES BETWEEN THE PLANES BELOW. EACH ANGLE MUST PASS THROUGH POINT O ON THE LINES OF INTERSECTION. RPRIMARY AUXILIARY VIEWS ARE REQUIRED TO SOLVE PROBLEMS 2 AND 3.
ANGLE:
3
1
A
C
B
O
D
H
F
C
B
A
D
2
1
D
A
E
O
C
B
F
H
F
C
D
F
B
E
A

# REVOLUTION – POINT ABOUT A LINE

**1** REVOLVE POINT O ABOUT LINE 1-2, SHOW THE PATH OF REVOLUTION IN ALL VIEWS, AND MARK THE HIGHEST POINT OF O ON THIS PATH.

**2** REVOLVE POINT O ABOUT AXIS AB LOCATING ITS MOST FORWARD POSITION (O'). DETERMINE THE RADIUS OF THE CIRCULAR PATH. SHOW O, O' AND THE PATH OF REVOLUTION IN ALL VIEWS.

RADIUS _____

1 2

O

H
F

+ 1

2

1

O

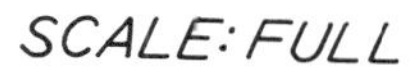


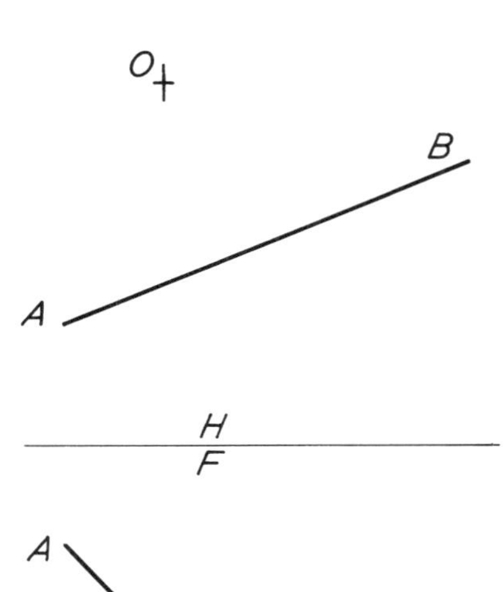


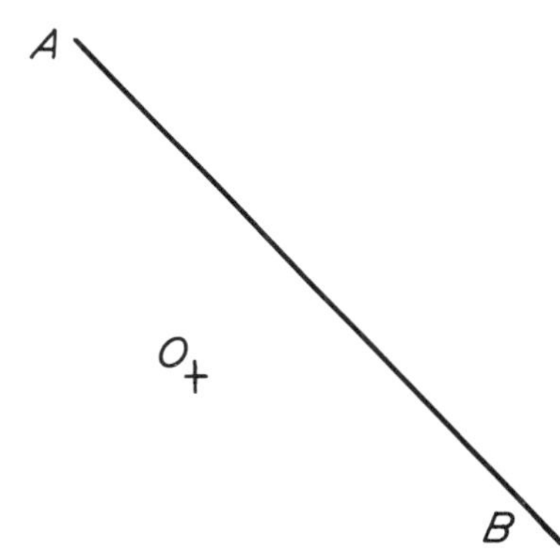


2 +

+ 1

**3** TO CHECK FOR ADEQUATE CLEARANCE OF THE HAND-OPERATED CRANK WITH EDGE AB, REVOLVE POINT O ABOUT THE AXIS AND SHOW THE PATH OF REVOLUTION IN ALL VIEWS. WHAT IS THE CLEARANCE?

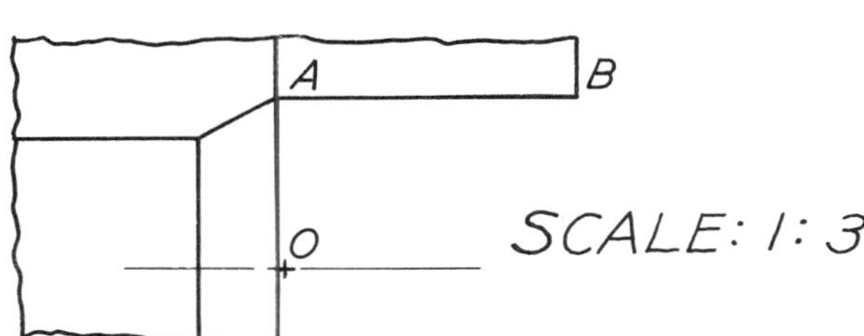


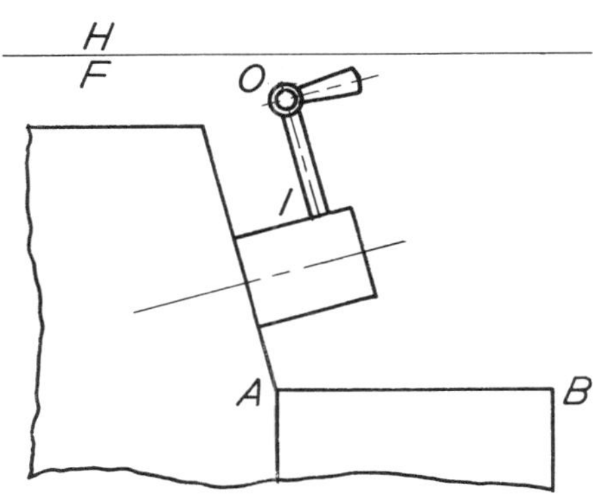


+ 1

CLEARANCE _____

## CONCURRENT, COPLANAR FORCES

THREE FORCES ACT UPON A GLIDER IN FLIGHT-- LIFT, DRAG, AND WEIGHT. FIND THE RESULTANT OF THIS SYSTEM OF FORCES SHOWN IN THE SKETCH.

A. SOLVE BY THE PARALLELOGRAM METHOD

B. SOLVE BY THE POLYGON METHOD

C. SOLVE BY THE MATHEMATICAL METHOD.

SCALE: 1 = 400 LBS

PARALLELOGRAM METHOD

POLYGON METHOD

R =

R =

MATHEMATICAL SOLUTION

$\Sigma F_X = R_X = (DRAG) \cos 45° - (LIFT) \cos 45°$

$=$

$R_X =$

$\Sigma F_Y = R_Y = (DRAG) \sin 45° + (LIFT) \sin 45° - WEIGHT$

$=$

$R_Y =$

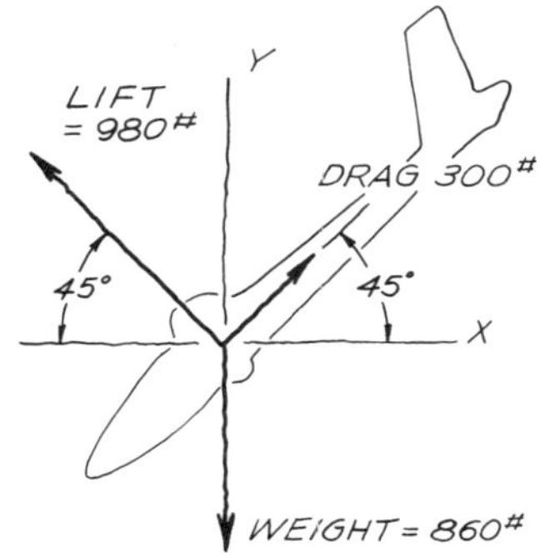


$\theta = ARCTAN \frac{R_Y}{R_X} =$

$\theta =$

$R = \sqrt{R_X^2 + R_Y^2} \quad =$

$R =$

# LETTERING GUIDE LINES

THESE GUIDELINES CAN BE USED TO UNDERLAY TRACING PAPER OR OTHER TRANSLUSCENT PAPERS WHEN LETTERING A DRAWING.

| Graphics for Engineers © | NAME<br>FILE SEC DATE | MIN. | GRADE | |
|---|---|---|---|---|

## CONCURRENT NONCOPLANAR VECTORS

DOUBLE THE ORTHOGRAPHIC PROJECTIONS OF THE FORCE SYSTEM GIVEN WHEN FINDING THE RESULTANT OF THE SYSTEM BY THE METHODS BELOW.

A. USING THE PARALLEL METHOD, ADD A AND B TO OBTAIN $R_1$; THEN ADD $R_1$ AND C.

B. USING THE POLYGON METHOD, LAY OUT THE VECTORS ALPHABETICALLY.

C. SOLVE MATHEMATICALLY. EQUATIONS 1 AND 2 REFER TO THE FRONT VIEW, AND EQUATION 3 TO THE TOP VIEW.

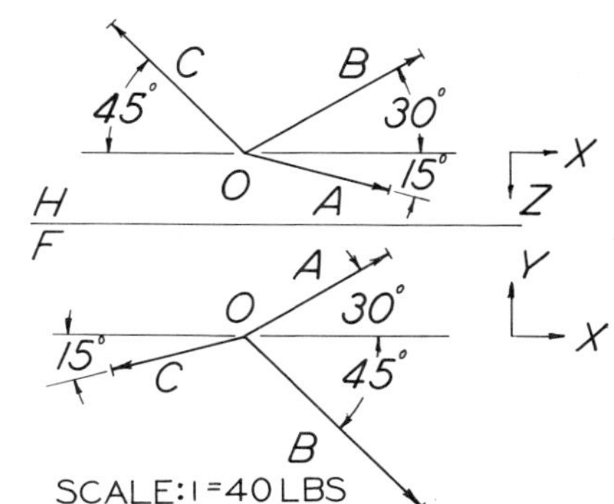


O

SCALE: 1 = 20 LBS

O

H PARALLELOGRAM METHOD

F

H POLYGON METHOD

F

O

C

R =

R =

TL DIAGRAM

TL DIAGRAM

1. $\Sigma F_X = R_X = A \cos 30° + B \cos 45° - C \cos 15°$

$R_X =$

2. $\Sigma F_Y = R_Y = A \sin 30° - B \sin 45° - C \sin 15°$

$R_Y =$

3. $\Sigma F_Z = R_Z = A \sin 15° - B \sin 30° - C \sin 45°$

$R_Z =$

4. $R = \sqrt{R_X^2 + R_Y^2 + R_Z^2} =$

| Graphics & Geometry © | NAME | MIN. | GRADE | 93 |
|---|---|---|---|---|
| | FILE SEC DATE | | | |

# LETTERING GUIDE LINES

THESE GUIDELINES CAN BE USED TO UNDERLAY TRACING PAPER OR OTHER TRANSLUSCENT PAPERS WHEN LETTERING A DRAWING.

| Graphics for Engineers © | NAME<br>FILE SEC DATE | MIN. | GRADE | |
|---|---|---|---|---|

VECTOR ANALYSIS

SCALE: 1=40#

**1** BY THE POLYGON METHOD, FIND THE LOADS CARRIED BY MEMBERS A AND B.

| MEMBER | LOAD | T or C |
|---|---|---|
| A | | |
| B | | |

BEGIN LOAD VECTOR HERE

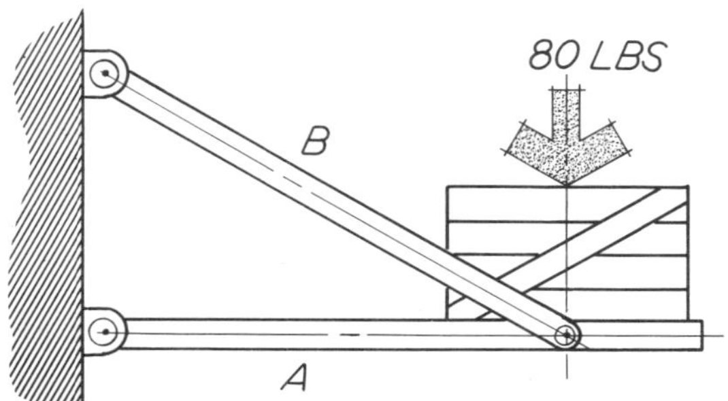

**2** DETERMINE THE LOADS IN MEMBERS A AND B. DRAW THE VECTOR POLYGON IN THE FOLLOWING ORDER: THE LOAD, THE CABLE, MEMBER A AND THEN MEMBER B.

BEGIN LOAD VECTOR HERE

SCALE: 1=200#

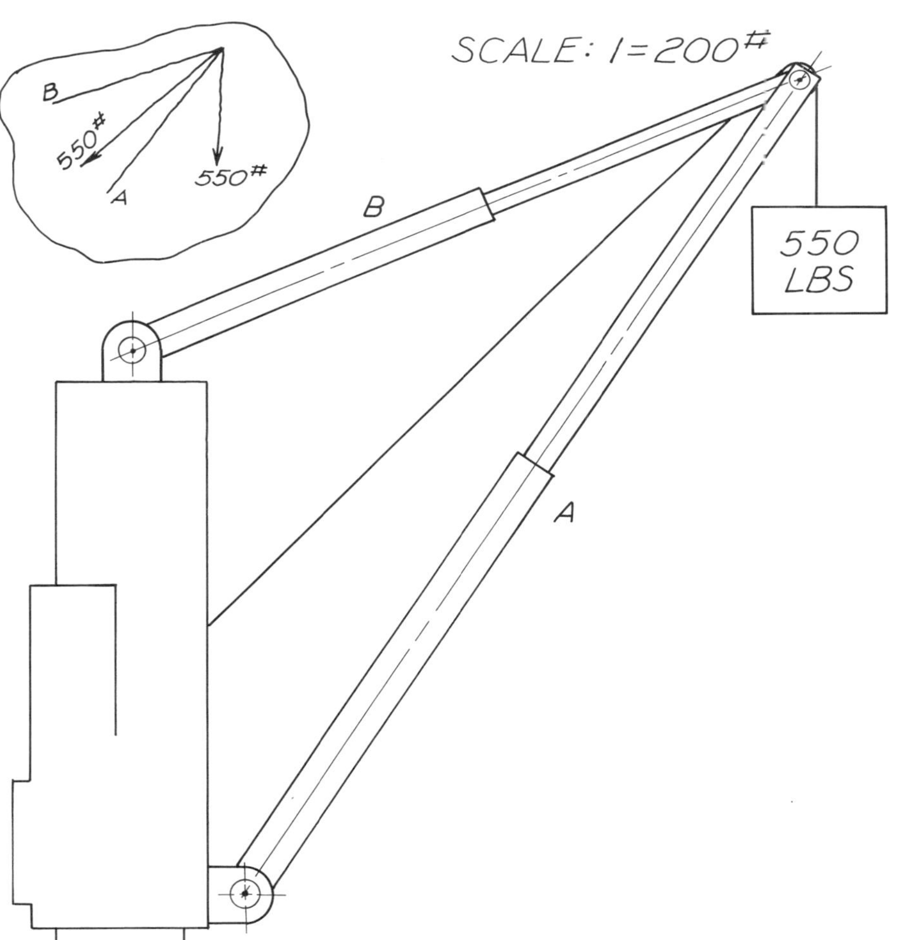


| MEMBER | LOAD | T or C |
|---|---|---|
| A | | |
| B | | |

# LETTERING GUIDE LINES

THESE GUIDELINES CAN BE USED TO UNDERLAY TRACING PAPER OR OTHER TRANSLUSCENT PAPERS WHEN LETTERING A DRAWING.

| Graphics for Engineers © | NAME<br>FILE SEC DATE | MIN. | GRADE | |
|---|---|---|---|---|

## TRUSS ANALYSIS

TO PROPERLY SIZE THE MEMBERS OF A TRUSS, THE LOADS THAT EACH WILL CARRY MUST BE DETERMINED.

A. BEGINNING WITH JOINT 1, DRAW THE VECTOR POLYGON FOR EACH JOINT.

B. CONSTRUCT A MAXWELL DIAGRAM. NOTICE THAT THIS DIAGRAM IS A SUPERPOSITION OF THE VECTOR POLYGONS OF EACH JOINT.

C. ON A SEPARATE SHEET, SOLVE MATHEMATICALLY.

5,000#
10,000#
8,000#
B
C
A
f
3
h
4
1
2
g
5
D
E
9,500#
13,500#

JOINT 1

A

JOINT 2

A

JOINT 3

B

JOINT 4

C

BEGIN MAXWELL DIAGRAM HERE

A

SCALE: 1=5,000#

JOINT 5

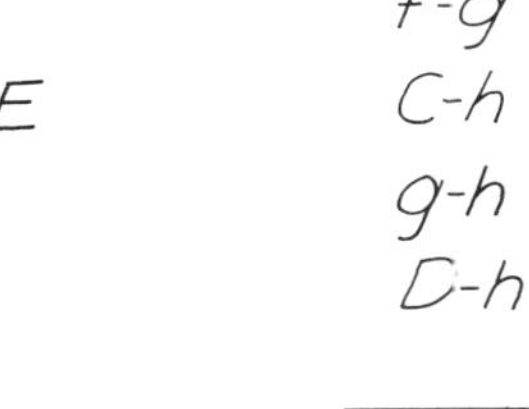

E

MEMBER LOAD CORT

f-A

B-f

E-g

f-g

C-h

g-h

D-h

| Graphics & Geometry © | NAME | MIN. | GRADE | 95 |
|---|---|---|---|---|
| | FILE SEC DATE | | | |

# LETTERING GUIDE LINES

THESE GUIDELINES CAN BE USED TO UNDERLAY TRACING PAPER OR OTHER TRANSLUSCENT PAPERS WHEN LETTERING A DRAWING.

| Graphics for Engineers © | NAME<br>FILE SEC DATE | MIN. | GRADE | |
|---|---|---|---|---|

## VECTOR ANALYSIS–SPECIAL CASE

IN THE ELECTRICAL CONSTRUCTION INDUSTRY WIRE IS NORMALLY PURCHASED ON REELS. DETERMINE THE LOADS IN LEGS OF THE REEL "JACK" SO THAT THE PROPER SIZE MAY BE SELECTED.

| MEMBER | LOAD | C OR T |
| --- | --- | --- |
| A | | |
| B | | |
| C | | |

COURTSEY OF C&M ELECTRIC

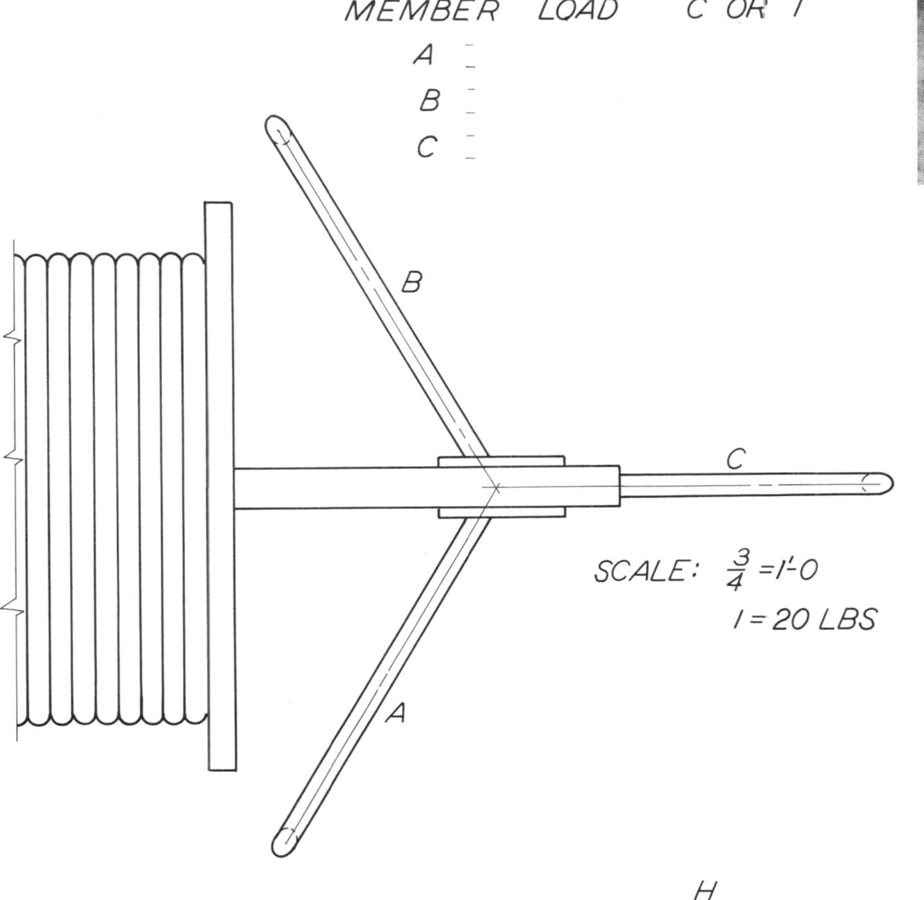


SCALE: $\frac{3}{4}$ = 1'-0

1 = 20 LBS

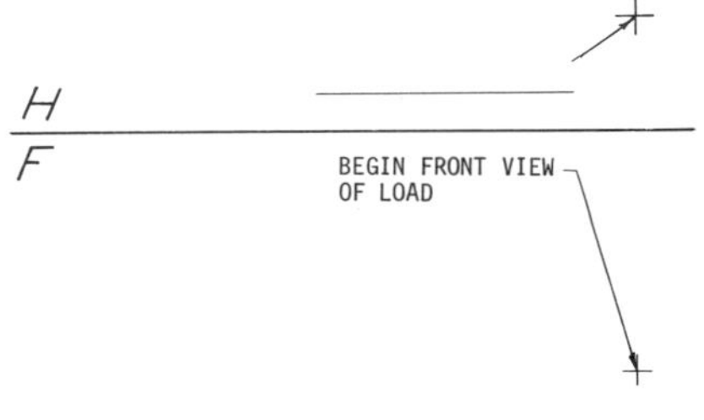


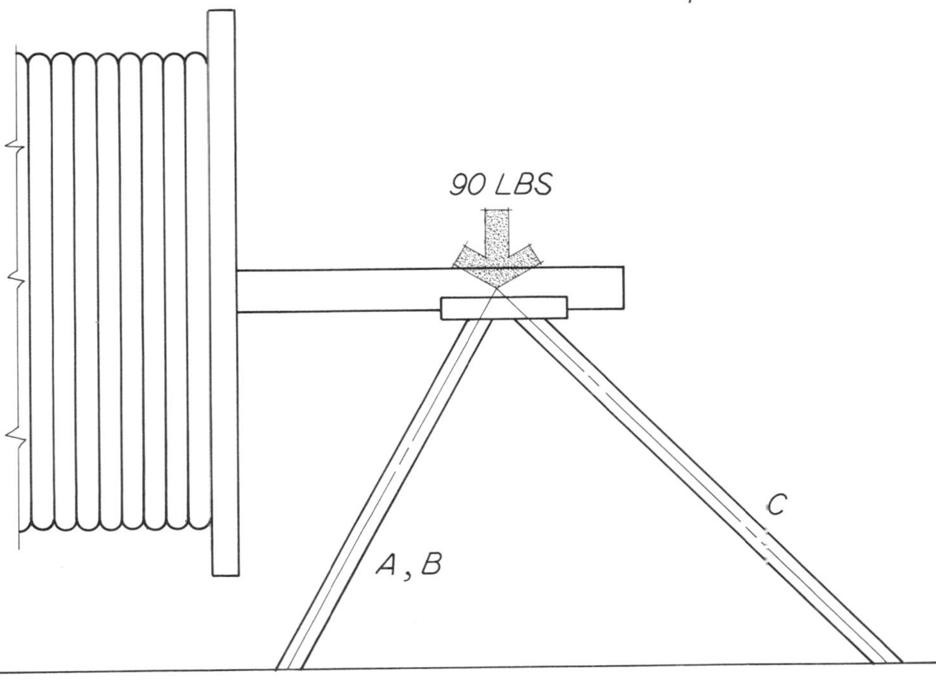

# LETTERING GUIDE LINES

THESE GUIDELINES CAN BE USED TO UNDERLAY TRACING PAPER OR OTHER TRANSLUSCENT PAPERS WHEN LETTERING A DRAWING.

| Graphics for Engineers © | NAME<br>FILE SEC DATE | MIN. | GRADE | |
|---|---|---|---|---|

## *NONCOPLANAR VECTORS IN EQULIBRIUM*

A 40 POUND FORCE IS APPLIED TO THE MIRCOWAVE APPARATUS. FIND THE LOADS IN MEMBERS A, B & C. BEGIN BY FINDING THE EDGE VIEW OF A PLANE THAT CONTAINS A AND B.

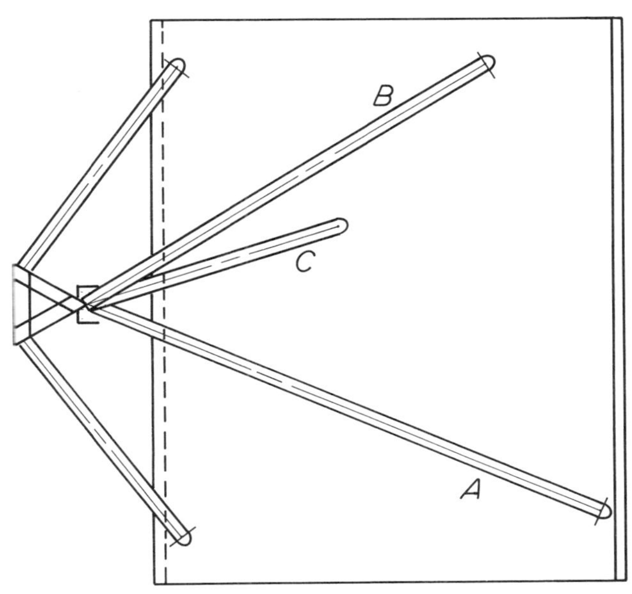


+ 1

Begin TL load

SCALE: 1 = 20 LBS
SCALE: 1 = 2'

40 LBS

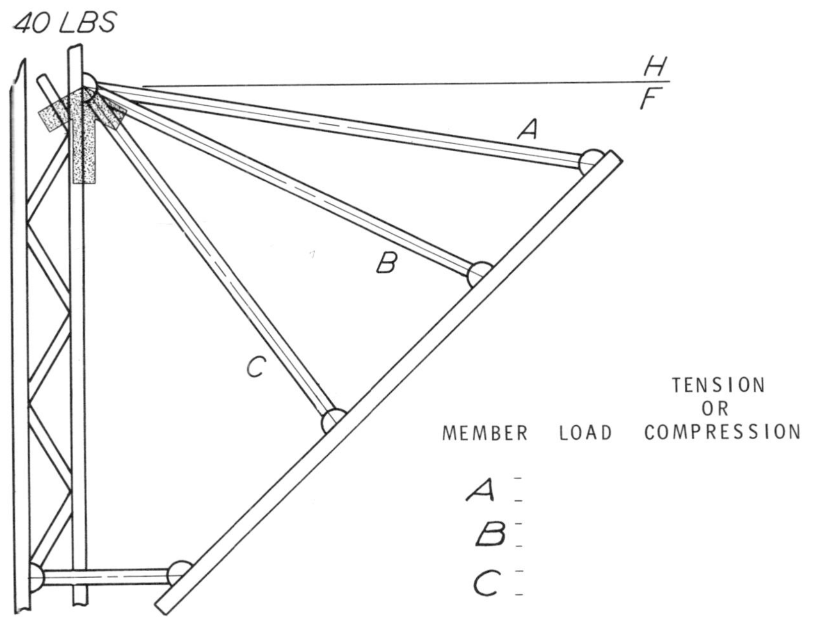


| MEMBER | LOAD | TENSION OR COMPRESSION |
|---|---|---|
| A | | |
| B | | |
| C | | |

COURTESY OF MICROFLECT

# LETTERING GUIDE LINES

THESE GUIDELINES CAN BE USED TO UNDERLAY TRACING PAPER OR OTHER TRANSLUSCENT PAPERS WHEN LETTERING A DRAWING.

| Graphics for Engineers © | NAME<br>FILE SEC DATE | MIN. | GRADE | |
|---|---|---|---|---|

## BEAM ANALYSIS

IN ORDER TO DETERMINE THE PROPER SIZE OF THE COLUMNS SUPPORTING THE BAR JOIST, THE LOADS CARRIED BY EACH MUST BE DETERMINED.

A. FIND THE LOADS BY THE GRAPHICAL METHOD.

B. FIND THE LOCATION ($\bar{X}$) OF A SINGLE EQUILIBRANT THAT COULD REPLACE THE TWO COLUMNS. MEASURE FROM THE LEFT END OF THE JOIST.

C. SOLVE A AND B BY THE MATHEMATICAL METHOD.

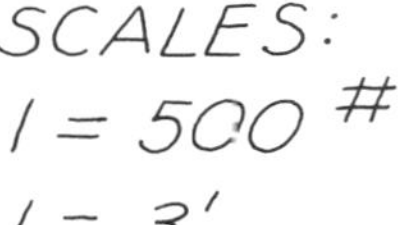

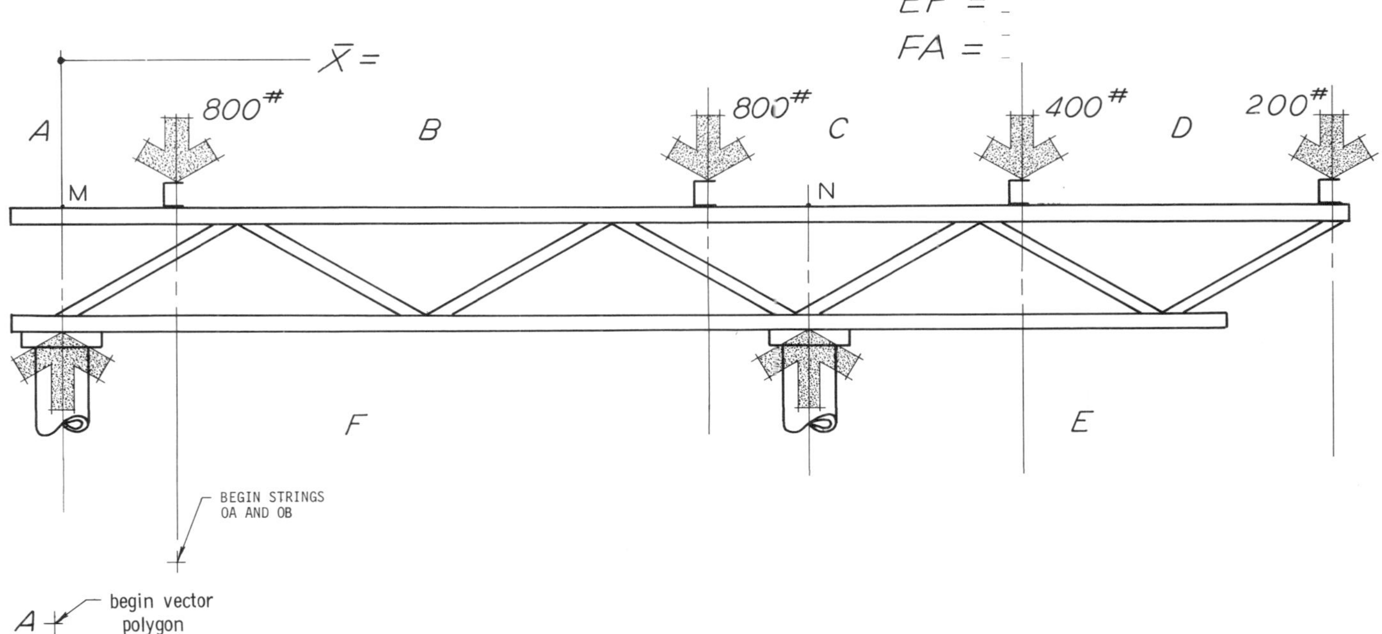


A — begin vector polygon

pole point for string diagram

$\Sigma M_M = 0 =$

EF =

$\Sigma F_V =$

# LETTERING GUIDE LINES

THESE GUIDELINES CAN BE USED TO UNDERLAY TRACING PAPER OR OTHER TRANSLUSCENT PAPERS WHEN LETTERING A DRAWING.

| Graphics for Engineers © | NAME<br>FILE SEC DATE | MIN. | GRADE | |
|---|---|---|---|---|

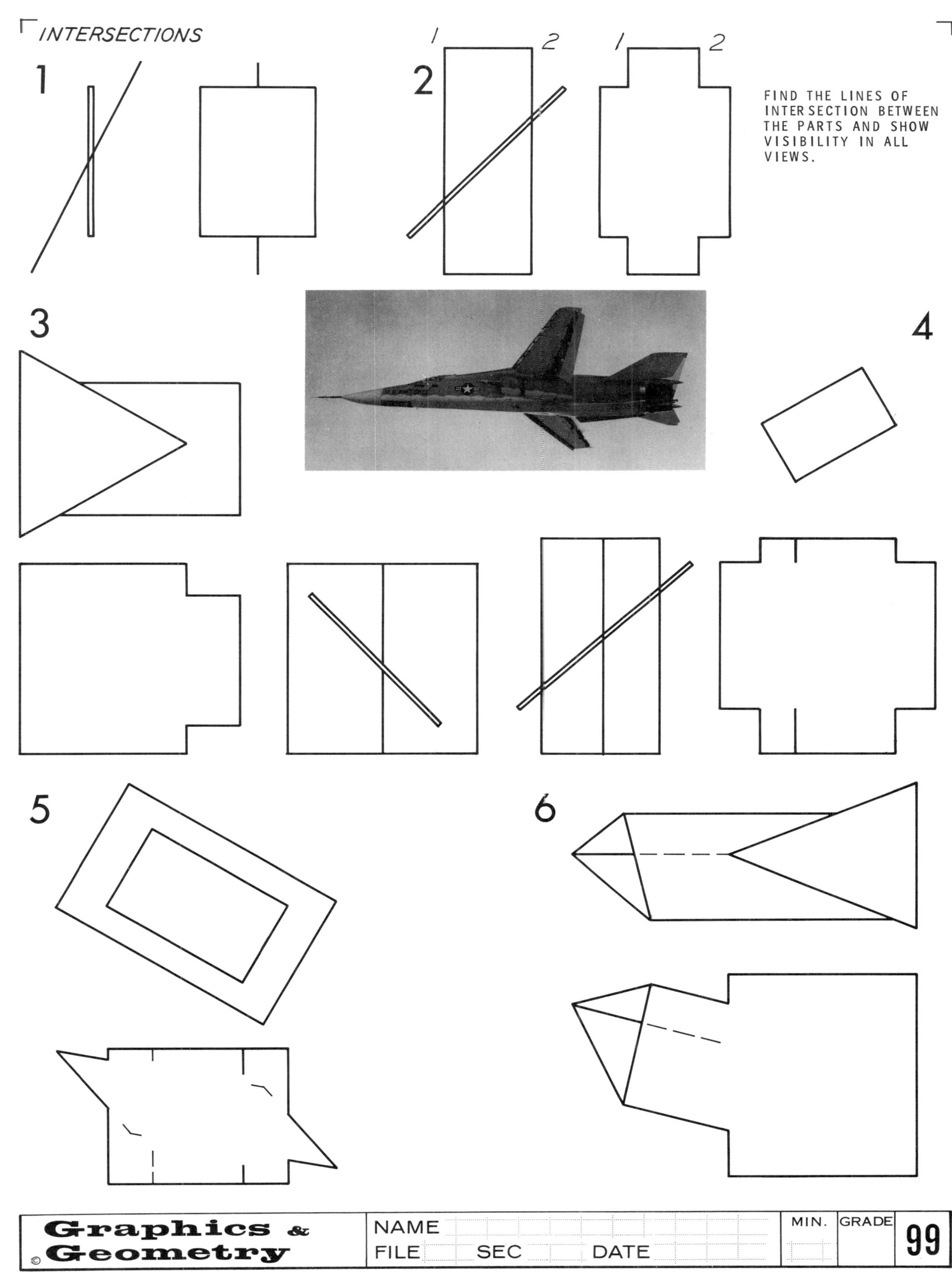
INTERSECTIONS
1
2
1
2
1
2
FIND THE LINES OF INTERSECTION BETWEEN THE PARTS AND SHOW VISIBILITY IN ALL VIEWS.
3
4
5
6

# LETTERING GUIDE LINES

THESE GUIDELINES CAN BE USED TO UNDERLAY TRACING PAPER OR OTHER TRANSLUSCENT PAPERS WHEN LETTERING A DRAWING.

| **Graphics for Engineers** © | NAME<br>FILE SEC DATE | MIN. | GRADE | |
|---|---|---|---|---|

FIND THE LINES OF INTERSECTION BETWEEN THE PARTS AND SHOW VISIBILITY IN ALL VIEWS.

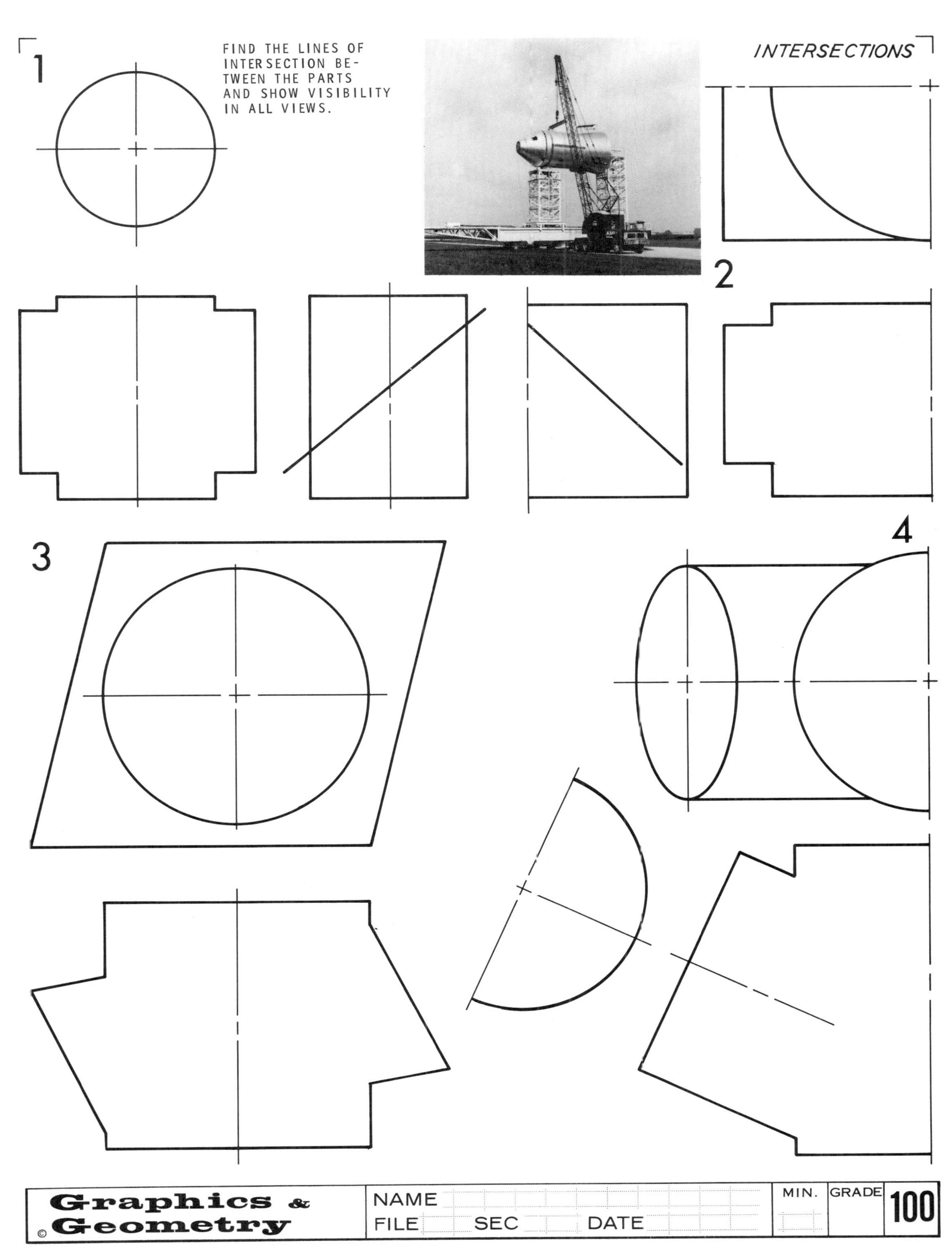

## INTERSECTIONS

FIND THE INTERSECTIONS BETWEEN THE SHEET METAL PARTS. SHOW VISIBILITY.

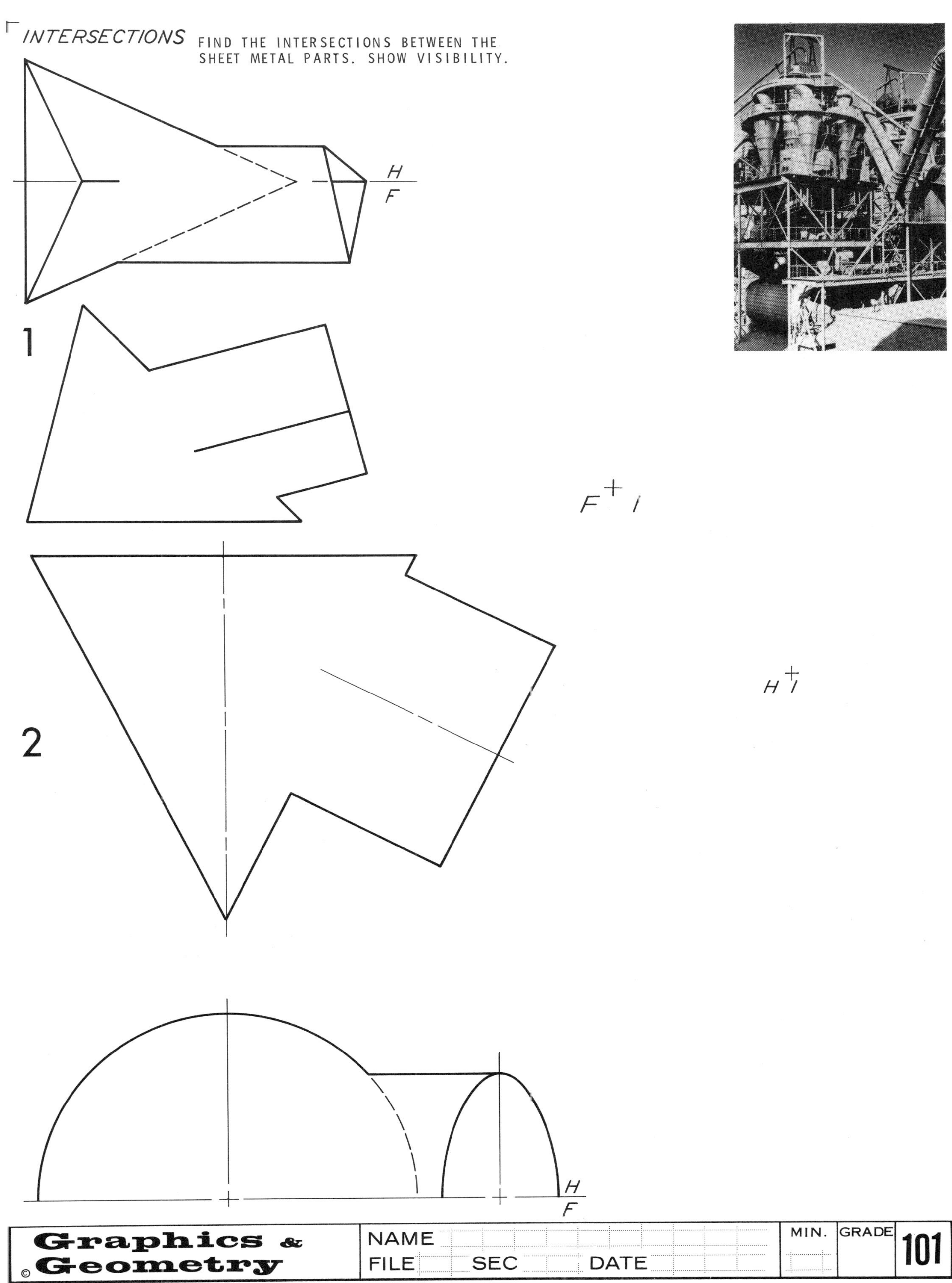

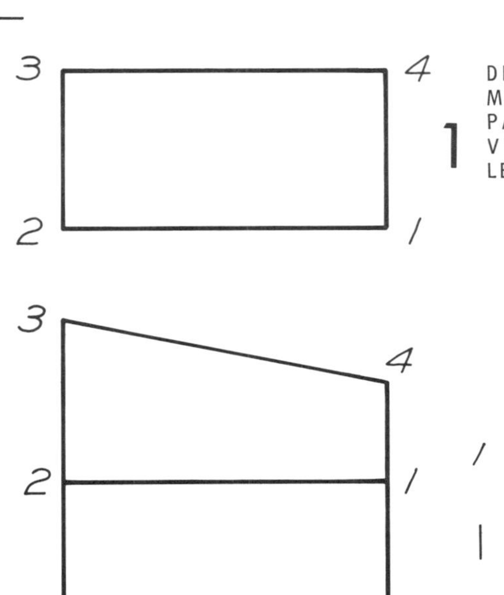


## DEVELOPMENTS

1 DEVELOP INSIDE PATTERNS OF THE SHEET METAL PARTS. BEGIN WITH THE SHORTEST PARTING LINES AND LABEL THE GIVEN VIEWS AND PATTERNS WITH NUMBERS OR LETTERS.

2

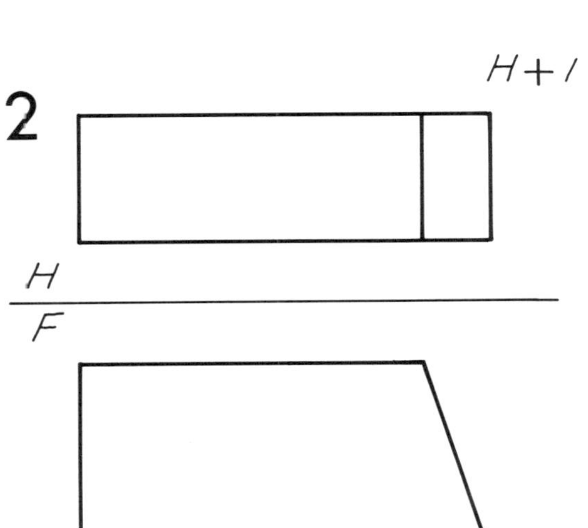


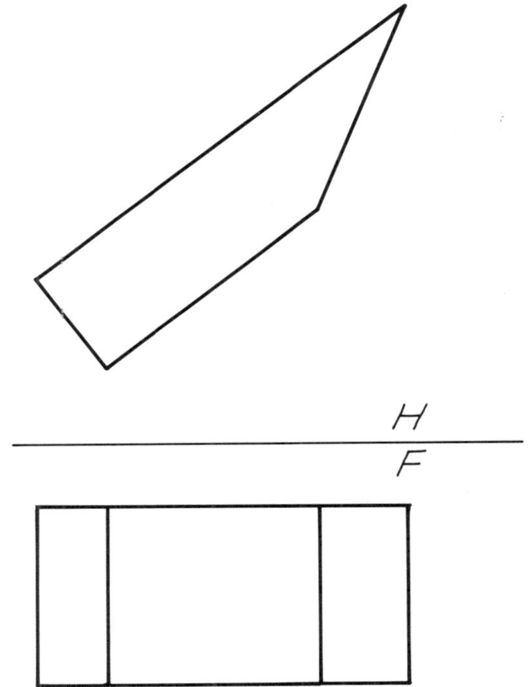

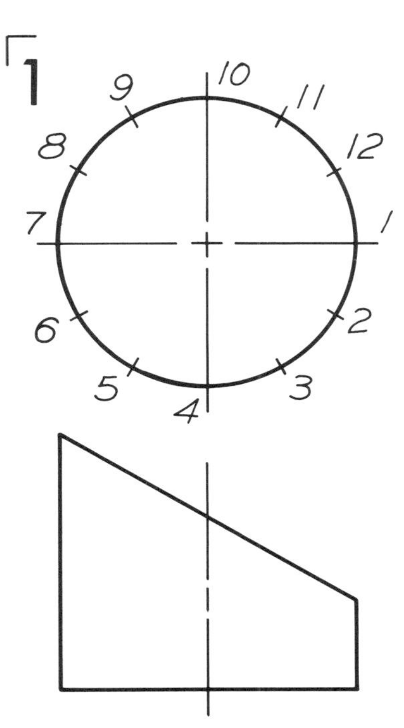


## DEVELOPMENTS

DEVELOP INSIDE PATTERNS OF THE SHEET METAL PARTS. BEGIN WITH THE SHORTEST PARTING LINES AND LABEL THE GIVEN VIEWS AND PATTERNS WITH NUMBERS OR LETTERS.

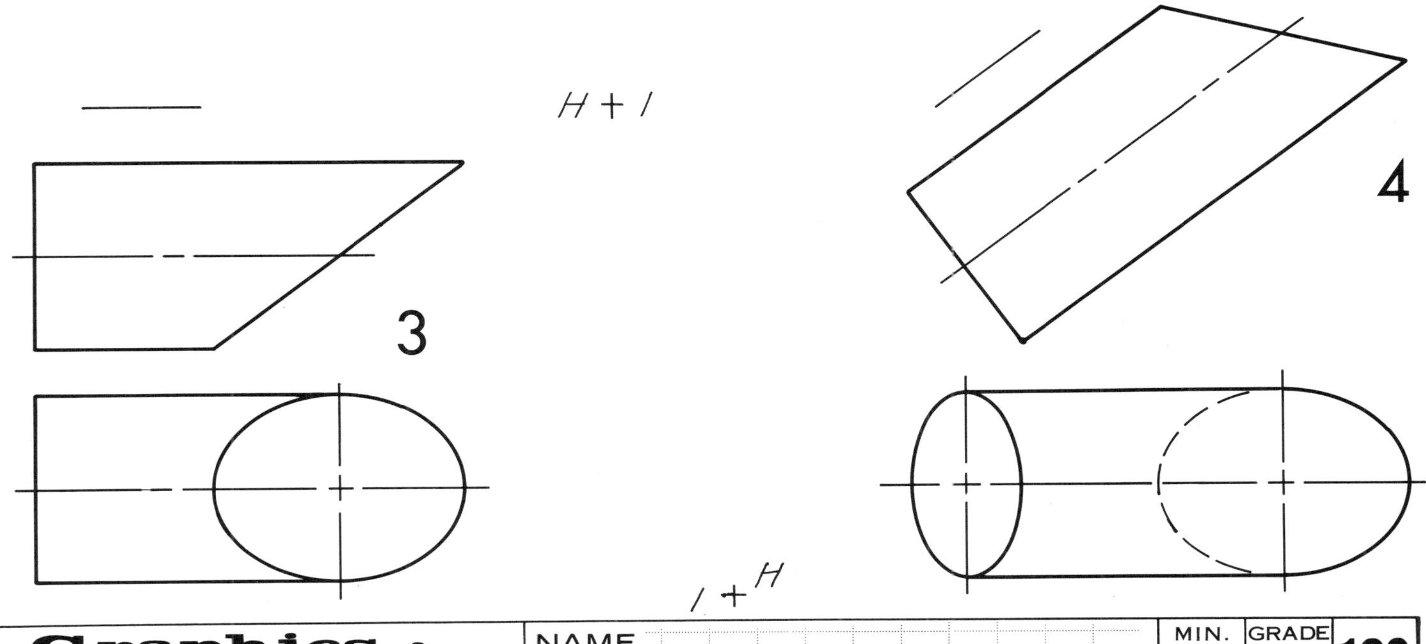


Graphics & Geometry © | NAME | FILE | SEC | DATE | MIN. | GRADE | 103

DEVELOPMENTS

1 LAY OUT AN INSIDE PATTERN OF THE PYRAMID. BEGIN WITN LINE 0-1 AND NUMBER THE POINTS.

2 DEVELOP AN INSIDE HALF PATTERN OF THE HOPPER FROM POINTS A TO D. BEGIN WITH LINE OA AND LETTER THE FOINTS.

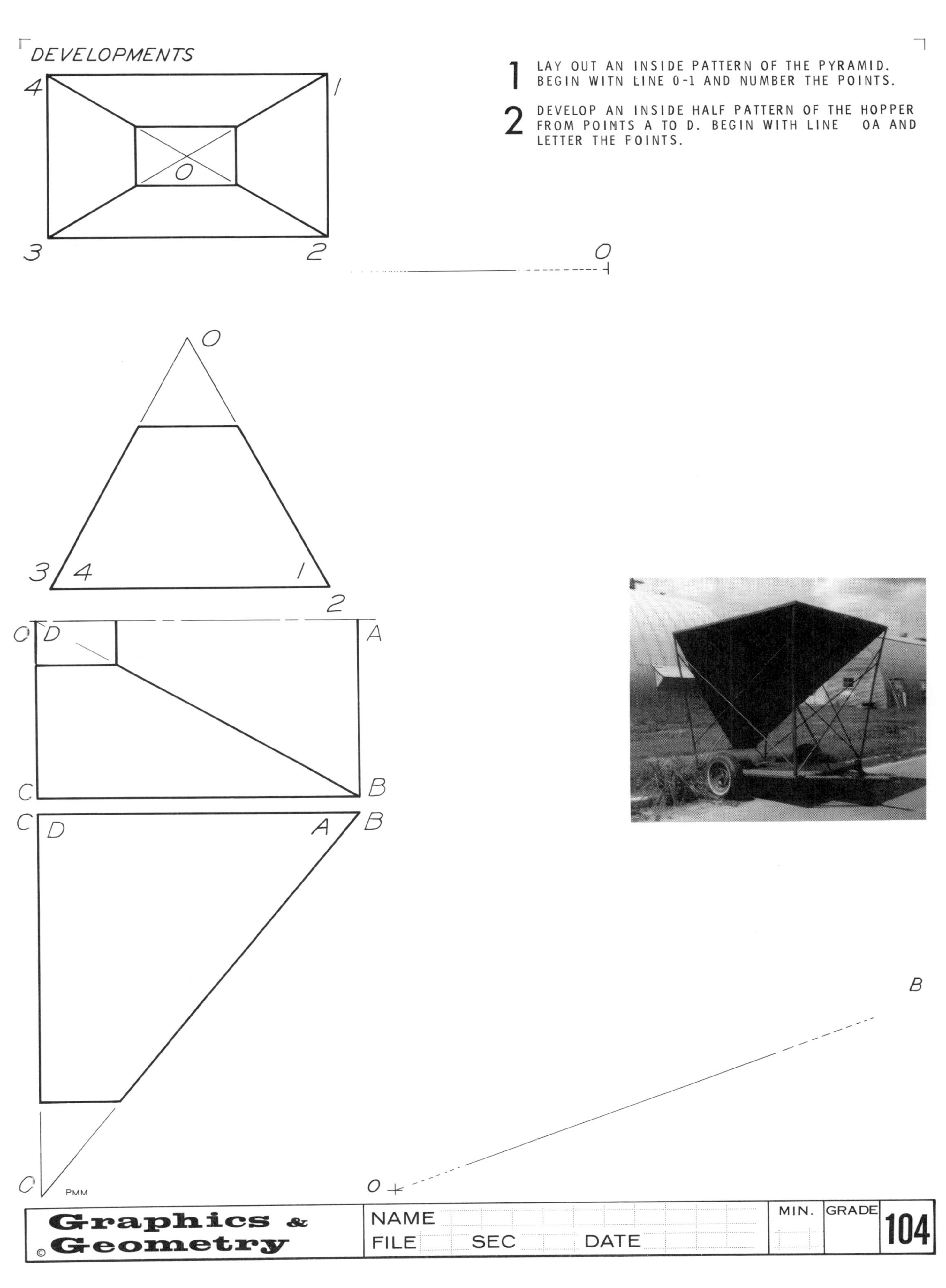

DEVELOPMENTS

LAY OUT INSIDE PATTERNS OF THE HALF-VIEWS OF THE SHEET METALS SHAPES BELOW. NUMBER THE POINTS. BEGIN WITH LINES 1-2 AND 0-1.

1

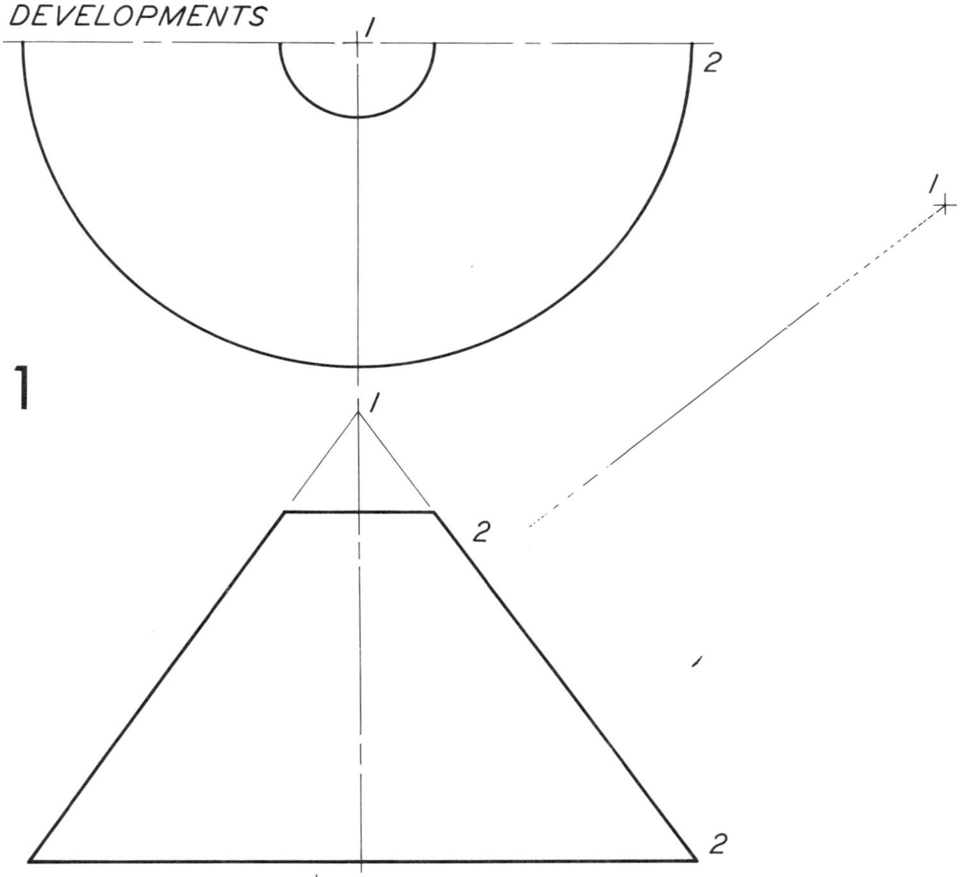

1 2 1 1 2 2

2

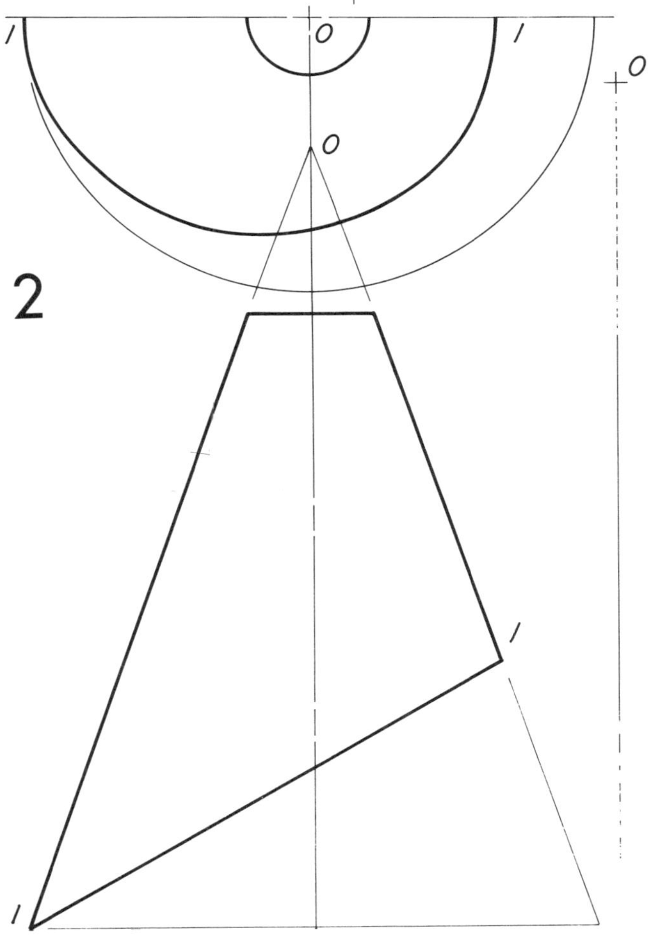

1 0 1 0 0 1 1

LAY OUT AN INSIDE FLAT PATTERN OF THE TRANSITION PIECE. BEGIN WITH LINE 1-2 THAT IS GIVEN. LABEL THE VIEWS AND THE PATTERN.

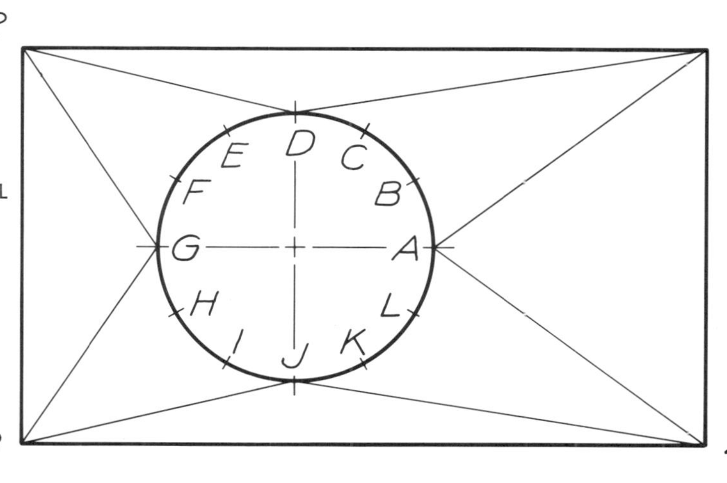

Kaiser Engineers

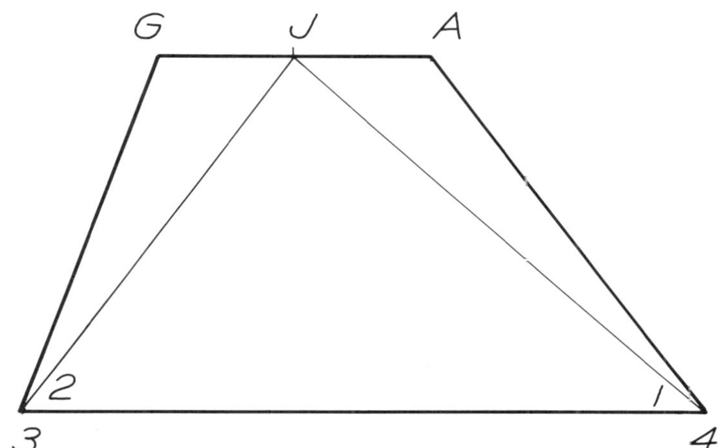


2 1

| CATEGORY | COST | % | ANGLE |
|---|---|---|---|
| RESEARCH | $20 | | |
| MATERIALS | 15 | | |
| LABOR | 40 | | |
| OVERHEAD | 12 | | |
| ADVERTISING | 25 | | |
| TOTALS | $112 | | |

THE FIGURES AT THE LEFT GIVE THE COSTS FOR MANUFACTURING A STEREO TAPE PLAYER BY THE AMERICAN SOUND COMPANY. CONVERT THESE FIGURES INTO A PIE GRAPH AND GIVE A TITLE UNDER THE GRAPH.

+

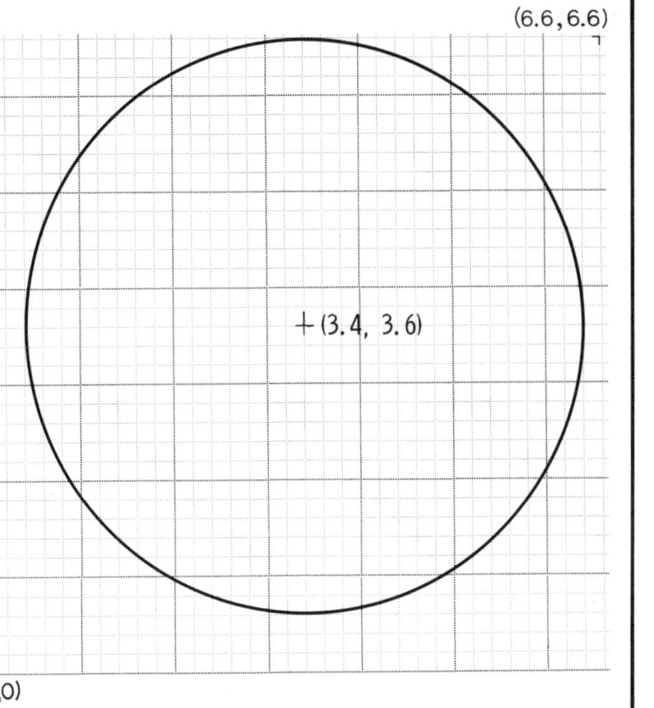


## CIRCLE GRAPH

The expenses for a single month for Smith Manufacturing Co. is itemized below. Plot this data in the form of a circle graph.

Smith Manufacturing Company

| | |
|---|---|
| Labor | $40,000 |
| Materials | 30,000 |
| Research | 10,000 |
| Management | 12,000 |
| Overhead | 8,000 |
| TOTAL | $100,000 |

# GRAPHS

IN CONSIDERING A DETERGENT TO BE USED IN PROPOSED CAR SERVICE CENTER, A STUDENT TEAM SKETCHED THE GRAPH AT THE RIGHT TO COMPARE FOUR DETERGENTS.

MAKE AND INSTRUMENT DRAWING OF THE BAR GRAPH THAT WOULD BE SUITABLE FOR USING IN A REPORT. DRAW IS TO BE AS BIG AS POSSIBLE IN THE SPACE BELOW AND HAVE PROPORTIONS OF 4:5.

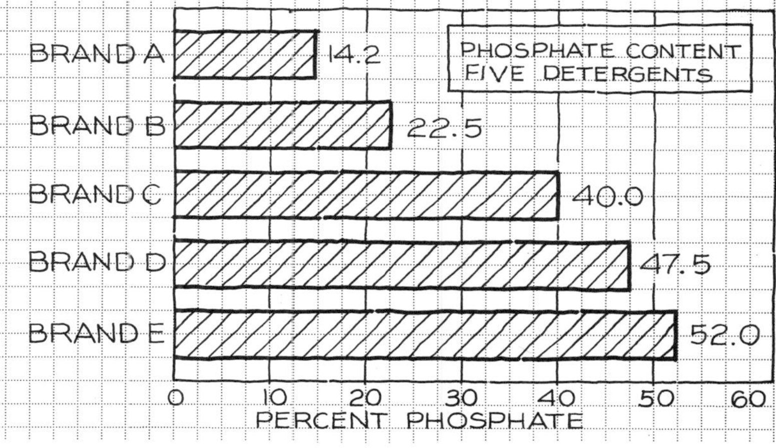

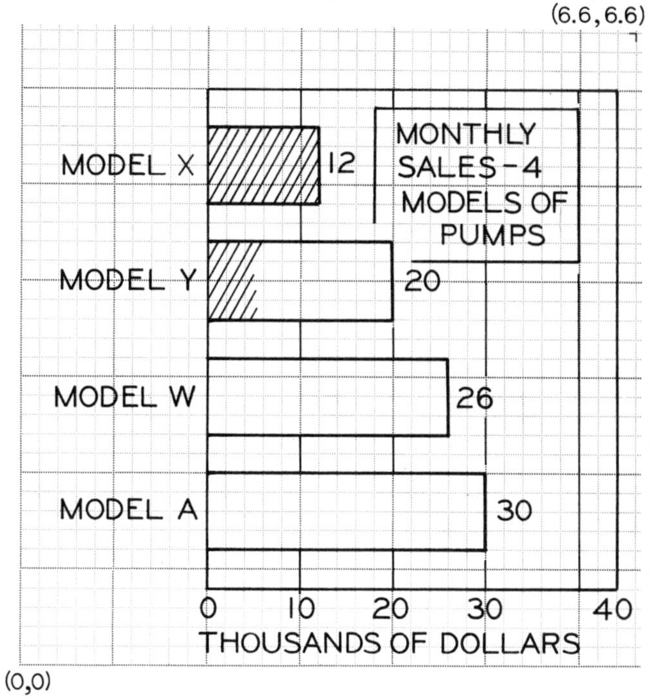

(6.6, 6.6)

MONTHLY SALES – 4 MODELS OF PUMPS

| Model | Thousands of dollars |
|---|---|
| MODEL X | 12 |
| MODEL Y | 20 |
| MODEL W | 26 |
| MODEL A | 30 |

0 10 20 30 40

THOUSANDS OF DOLLARS

(0,0)

## BAR GRAPH

Construct the bar graph above in the space at the left. Hatch all bars to match those that are hatched.

Position the title box in a better position so it will not overlap a bar. Use 0.125" letters.

# GRAPHS

A DESIGN TEAM INVESTIGATED THE FEASIBILITY OF LOCATING A CHILD CARE CENTER IN A SHOPPING CENTER. THEY COUNTED CHILDREN IN THE MALL AT VARIOUS TIMES AND SKETCHED THIS GRAPH TO SHOW THEIR FINDINGS.

CONVERT THIS ROUGH SKETCH INTO A FINISHED GRAPH USING INSTRUMENTS. DRAW THE GRAPH AS LARGE AS POSSIBLE AT A 4:5 RATIO IN THE AVAILABLE SPACE.

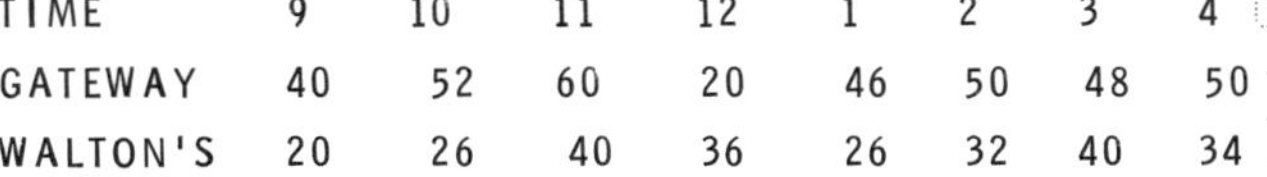

| TIME | 9 | 10 | 11 | 12 | 1 | 2 | 3 | 4 |
|---|---|---|---|---|---|---|---|---|
| GATEWAY | 40 | 52 | 60 | 20 | 46 | 50 | 48 | 50 |
| WALTON'S | 20 | 26 | 40 | 36 | 26 | 32 | 40 | 34 |

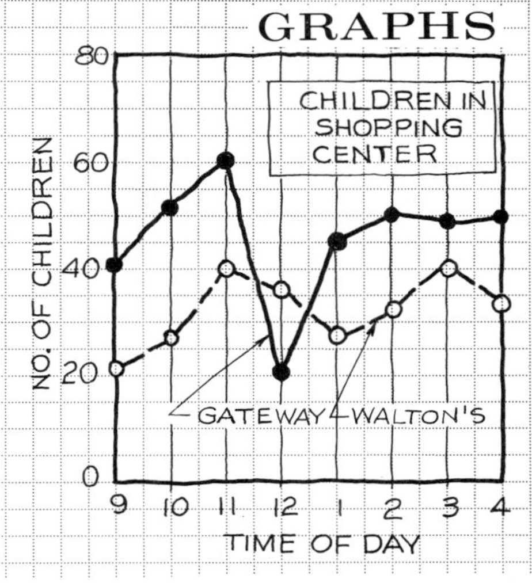

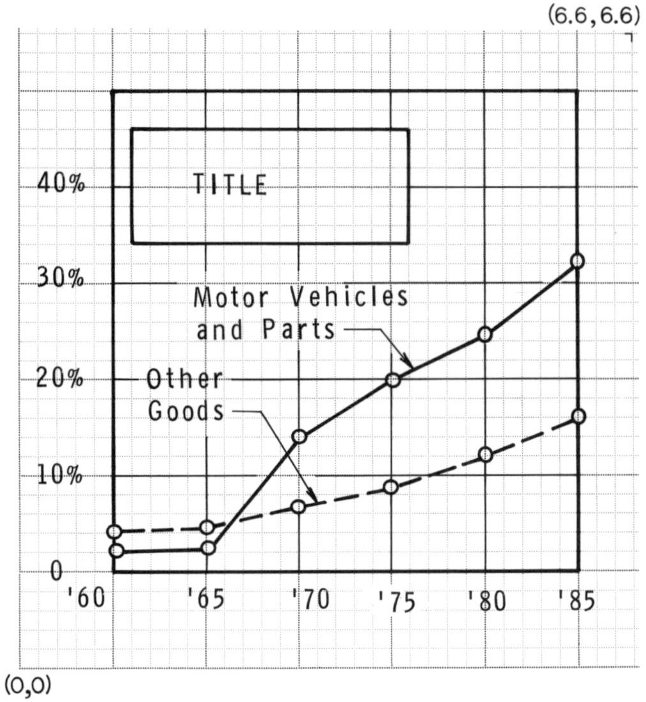

(6.6, 6.6)

40%
30%
20%
10%
0
'60 '65 '70 '75 '80 '85
TITLE
Motor Vehicles and Parts
Other Goods

(0,0)

GRAPH

The graph above compares the percent of import of motor vehicles and parts versus the import of other goods from foreign countries.

Construct this graph, label the axes, and provide a title.

## BREAK-EVEN GRAPH

CONSTRUCT A BREAK-EVEN GRAPH FOR THE MANUFACTURE OF A NEW PRODUCT THAT HAS A RESEARCH AND DEVELOPMENT COST OF $24,000 AND COSTS $2.00 EACH TO MANUFACTURE. IF YOU WISH TO BREAK EVEN WHEN 8,000 ARE SOLD, WHAT MUST BE CHARGED FOR THE PRODUCT? PLOT THIS DATA AND ANSWER THE QUESTIONS AT THE RIGHT.

1. PRICE OF EACH REQUIRED TO BREAK EVEN AT 8,000
2. LOSS IF ONLY 5,000 ARE SOLD
3. PROFIT IF 16,000 ARE SOLD

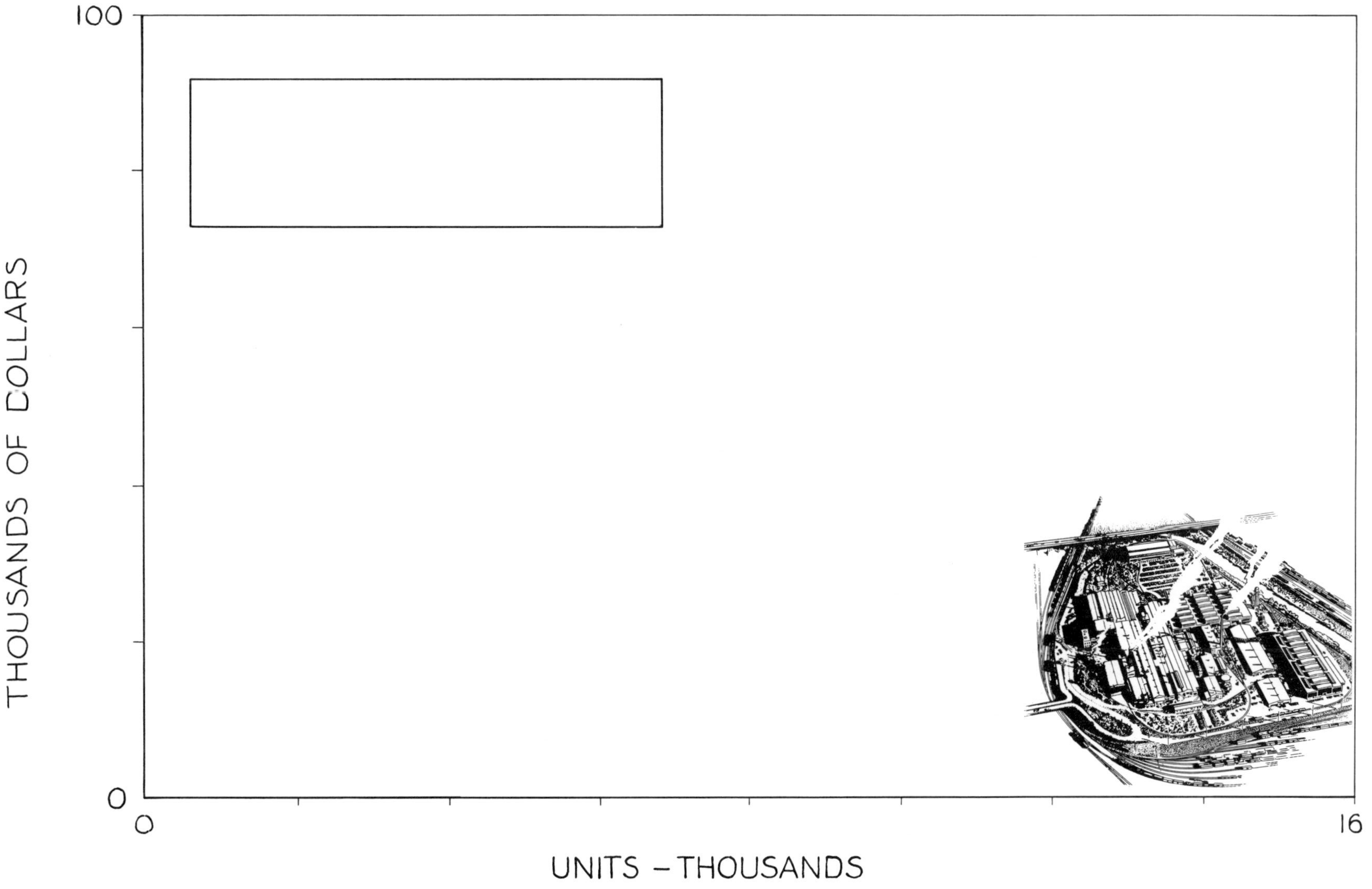

100

0

THOUSANDS OF DOLLARS

0 16

UNITS – THOUSANDS

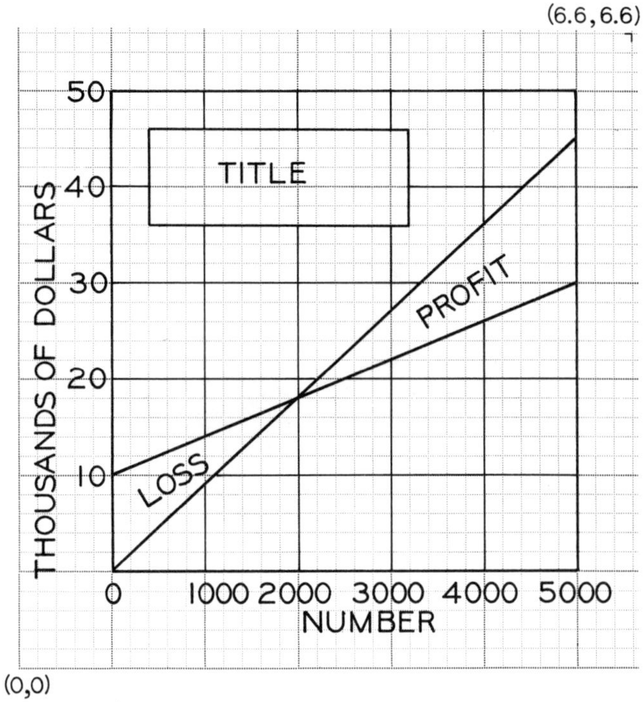

(6.6, 6.6)

| THOUSANDS OF DOLLARS | | | | | | |
|---|---|---|---|---|---|---|
| 50 | | | | | | |
| 40 | | TITLE | | | | |
| 30 | | | | PROFIT | | |
| 20 | | | | | | |
| 10 | LOSS | | | | | |
| | 0 | 1000 | 2000 | 3000 | 4000 | 5000 |
| | NUMBER | | | | | |

(0,0)

BREAK-EVEN GRAPH

Construct a break-even graph of the data below using the principles of graph construction.

A Zippo Calculator is to be modified with a development cost of $10,000. It will cost $4 per unit to manufacture them. Plot a graph that shows the break-even point at 2,000 units sold.

# GRAPHS

THIS FIELD DATA COMPARES THE ACCEPTANCE (IN PERCENT) OF AUTOMOBILES ONTO A HIGHWAY FROM ENTRY RAMPS OF 700 FT AND 1000 FT FOR VARIOUS GAPS (IN SECONDS) IN THE FLOW OF TRAFFIC. PLOT THIS DATA AND DRAW THE BEST CURVES USING THE EXAMPLES SHOWN. LETTER AND LABEL EACH AXIS AND PROVIDE A TITLE.

| PERCENT ACCEPTANCE FOR ENTRANCE RAMP OF LENGTH | GAP-SECONDS | | | | | | | | |
|---|---|---|---|---|---|---|---|---|---|
| | 1.0 | 1.5 | 2.0 | 3.0 | 4.0 | 5.0 | 6.0 | 7.0 | 8.0 |
| 1000 FT | 51 | 62 | 68 | 76 | 82 | 88 | 84 | 87 | 90 |
| | 60 | 66 | 76 | 80 | 82 | 81 | 90 | 90 | 90 |
| 700 FT | 22 | 37 | 40 | 45 | 56 | 60 | 67 | 73 | 73 |
| | 28 | 32 | 44 | 50 | 60 | 67 | 67 | 67 | 73 |

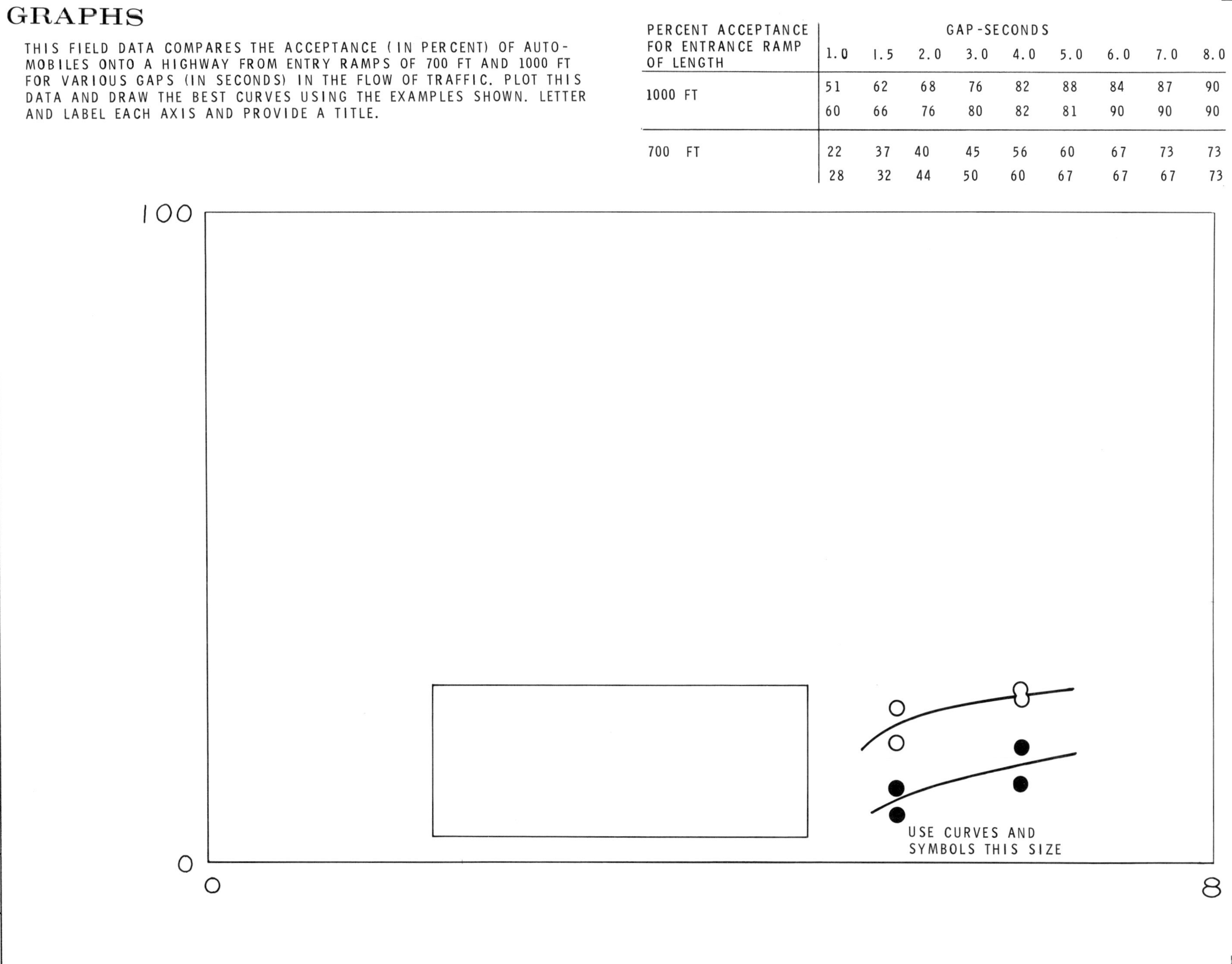

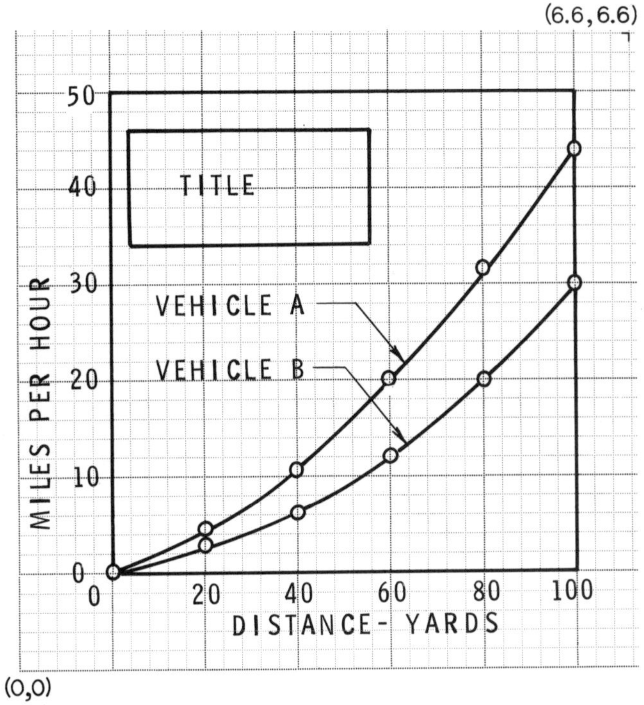

(6.6, 6.6)

TITLE

VEHICLE A

VEHICLE B

MILES PER HOUR

50
40
30
20
10
0

0 20 40 60 80 100

DISTANCE- YARDS

(0,0)

GRAPH

The graph above compares the acceleration of two vehicles. Construct this graph, provide a title, a draw the necessary index lines to make the graph more readable.

## GRAPHICAL ANALYSIS

SEMI-LOG GRAPHS ARE CALLED RATIO GRAPHS SINCE RATIOS OF VALUES CAN BE READ FROM THEM WITHOUT MATHEMATICAL COMPUTATIONS. DATA FOR TWO TYPES OF CONCRETE (TYPE 1 AND TYPE 3) ARE GIVEN IN THE TABLE. PLOT THIS DATA AND ANSWER THE QUESTIONS BELOW.

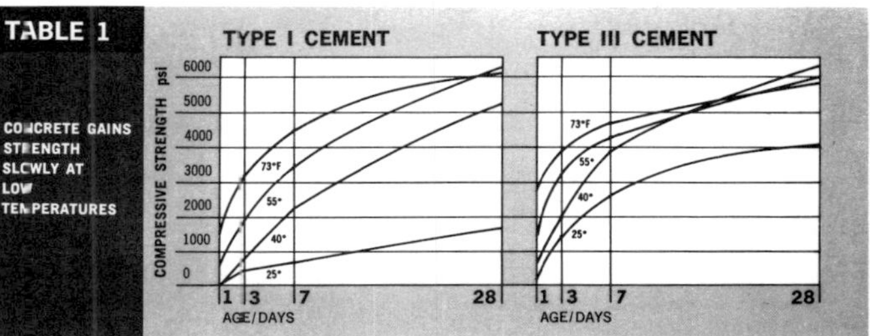


COURTESY OF KAISER CEMENT

1. THE STRENGTH OF TYPE 1 CONCRETE IS WHAT PERCENT OF TYPE 3 AT:

   4 DAYS? ___

   28 DAYS? ___

2. WHAT IS THE PERCENT OF INCREASE IN STRENGTH OF TYPE 3 OVER TYPE 1 AT:

   8 DAYS? ___

   28 DAYS? ___

COMPRESSIVE STRENGTHS OF TWO TYPES OF CONCRETE @ 25°F.

| DAYS OF AGING | 0 | 4 | 8 | 12 | 16 | 20 | 24 | 28 |
|---|---|---|---|---|---|---|---|---|
| TYPE 1 | 0 | 520 | 710 | 980 | 1220 | 1580 | 1620 | 1790 |
| TYPE 3 | 0 | 1610 | 2760 | 3250 | 3600 | 3810 | 3900 | 4010 |

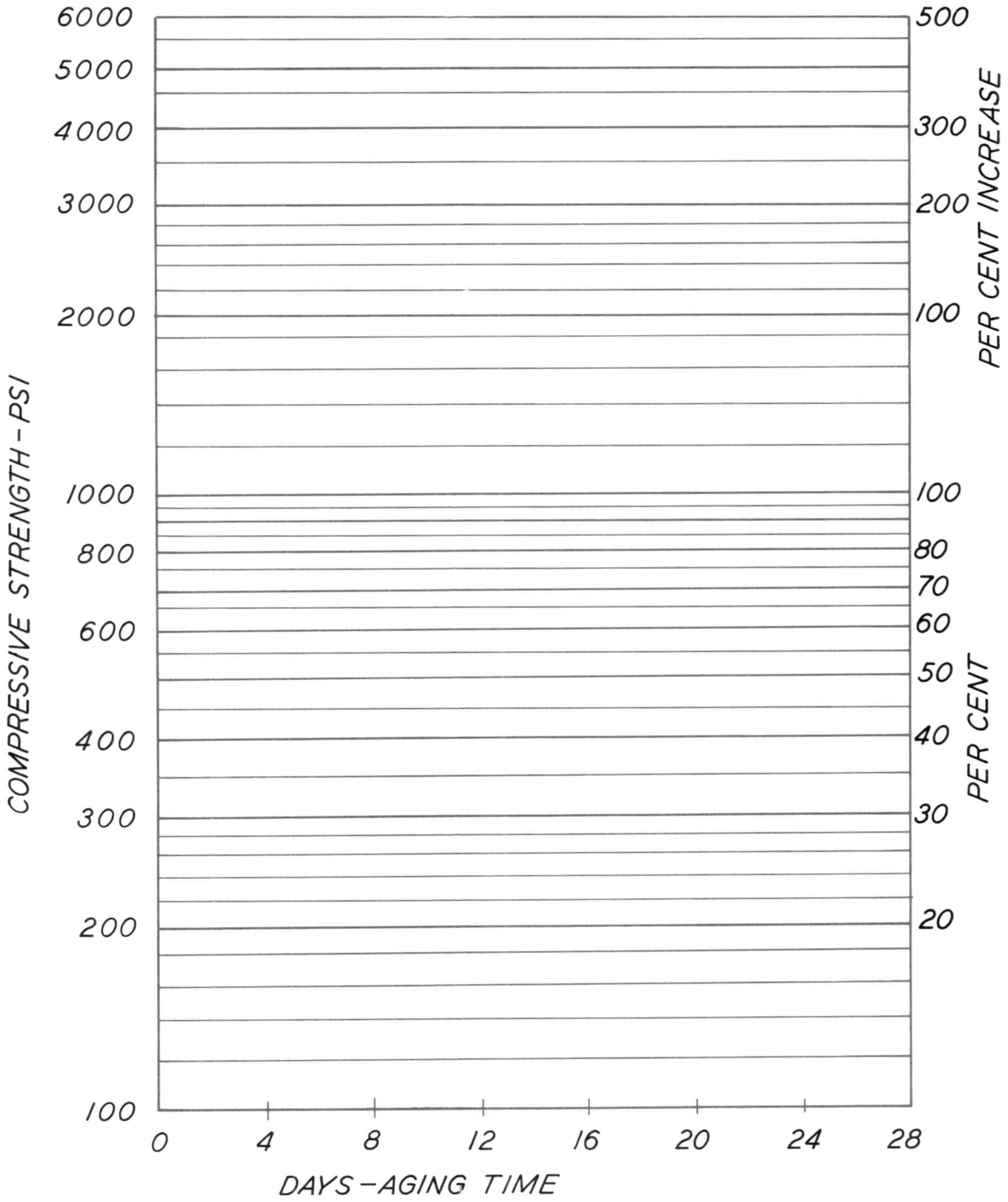

COMPRESSIVE STRENGTH-PSI: 6000, 5000, 4000, 3000, 2000, 1000, 800, 600, 400, 300, 200, 100

PER CENT INCREASE: 500, 300, 200, 100

PER CENT: 100, 80, 70, 60, 50, 40, 30, 20

DAYS-AGING TIME: 0, 4, 8, 12, 16, 20, 24, 28

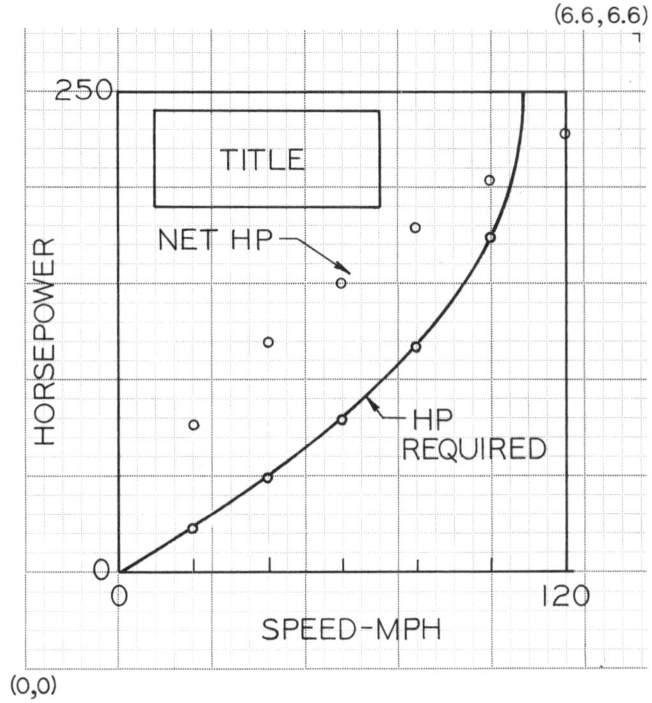

(6.6, 6.6)

250

TITLE

NET HP

HORSEPOWER

HP REQUIRED

0

0

120

SPEED-MPH

(0,0)

## DATA ANALYSIS

Data for two curves for an experimental car, Car X, are shown on the partially completed graph. One curve gives the horsepower required at various speeds, and the other shows the net horsepower that is available.

Plot the data and complete the graph using principles of good construction. Provide a title.

## *DATA ANALYSIS*

A WINDTUNNEL TEST HAS BEEN RUN ON AN AIRFOIL AND THE FOLLOWING MANOMETER READINGS HAVE BEEN RECORDED IN THE TABLES BELOW.

1. PLOT BOTH CURVES--STATIC PRESSURE ON UPPER SURFACE AND STATIC PRESSURE ON LOWER SURFACE--FOR EACH VALUE OF X (DISTANCE FROM THE TRAILING EDGE.)
2. GRAPHICALLY SUBTRACT THE TWO CURVES AND PLOT THIS DIFFERENCE USING THE ORDINATE SCALE AT THE RIGHT OF THE GRAPH.
3. ON A SEPARATE SHEET, GRAPHICALLY INTEGRATE THE DIFFERENCE CURVE TO FIND THE DIFFERENCE PER FOOT.

QUESTIONS

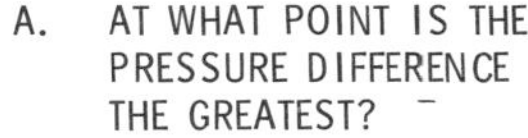

A. AT WHAT POINT IS THE PRESSURE DIFFERENCE THE GREATEST? _

B. AT WHAT POINT IS THE UPPER SURFACE PRESSURE THE LEAST? _

THE GREATEST? _

C. AT WHAT POINT IS THE LOWER SURFACE PRESSURE THE LEAST? _

THE GREATEST? _

COURTESY REPUBLIC AVIATION CORPORATION

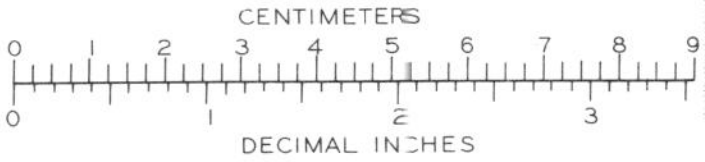

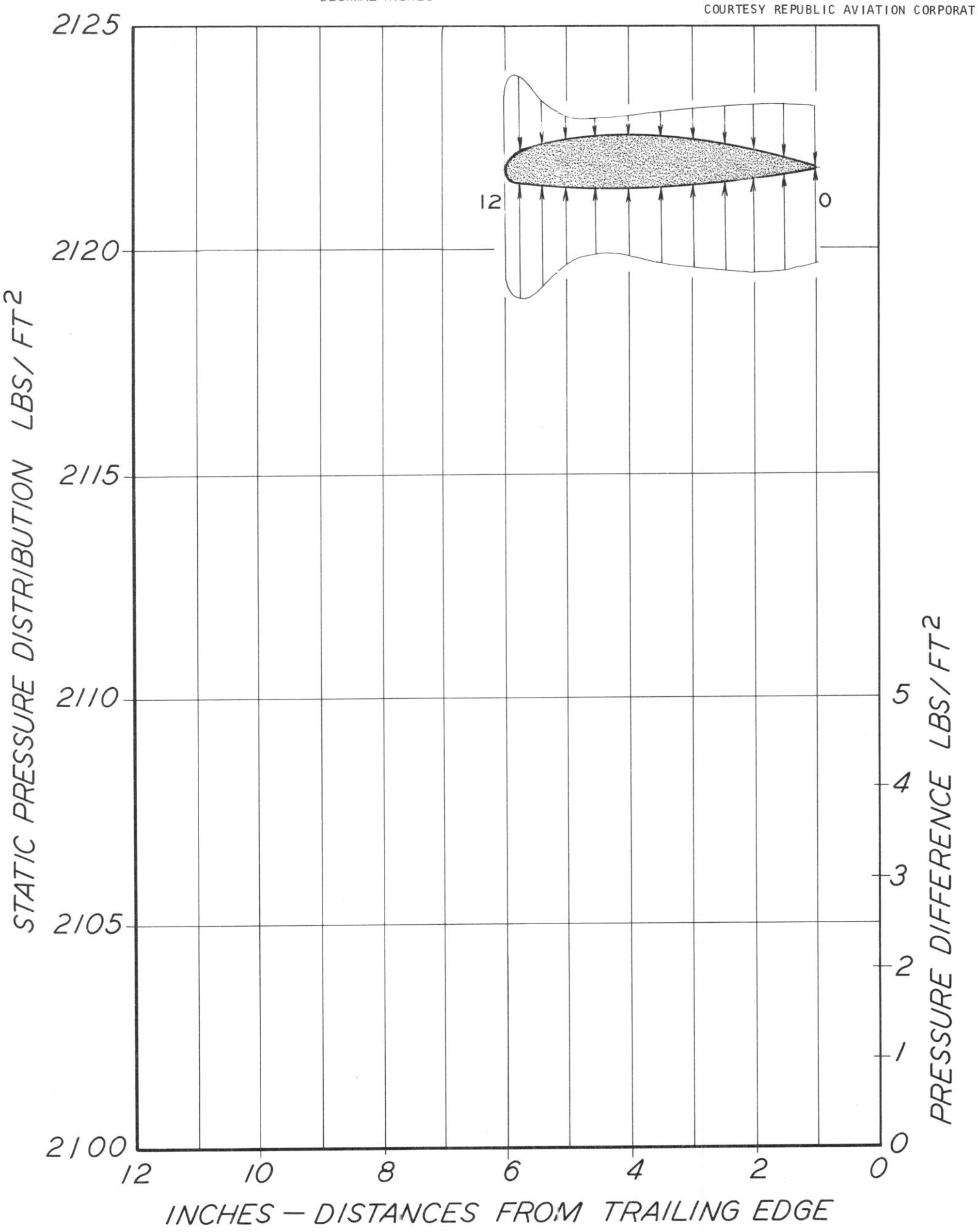

| DISTANCE FROM TRAILING EDGE | STATIC PRESSURE UPPER SURFACE LBS/SQ FT | STATIC PRESSURE LOWER SURFACE LBS/SQ FT |
|---|---|---|
| 12 | 2123.8 | 2123.8 |
| 11 | 2110.0 | 2113.2 |
| 10 | 2109.2 | 2113.2 |
| 9 | 2110.8 | 2113.3 |
| 8 | 2111.8 | 2113.4 |
| 7 | 2112.4 | 2113.8 |
| 6 | 2113.0 | 2114.0 |
| 5 | 2113.8 | 2114.4 |
| 4 | 2114.2 | 2115.0 |
| 3 | 2114.8 | 2115.6 |
| 2 | 2115.2 | 2116.2 |
| 1 | 2116.2 | 2116.4 |

STATIC PRESSURE DISTRIBUTION LBS/ FT²

PRESSURE DIFFERENCE LBS/ FT²

INCHES — DISTANCES FROM TRAILING EDGE

**Graphics & Geometry** ©

NAME

FILE SEC DATE

MIN. GRADE

# LETTERING GUIDE LINES

THESE GUIDELINES CAN BE USED TO UNDERLAY TRACING PAPER OR OTHER TRANSLUSCENT PAPERS WHEN LETTERING A DRAWING.

| Graphics for Engineers © | NAME | | MIN. | GRADE | |
|---|---|---|---|---|---|
| | FILE | SEC | DATE | | |

## *NOMOGRAM*

A NOMOGRAM CAN BE DESIGNED TO COMPUTE FINAL GRADES BASED ON COURSE AVERAGE AND FINAL EXAMINATION. COMPLETE THE PARALLEL SCALE CHART BELOW TO DETERMINE FINAL GRADES BASED ON THE FORMULA BELOW:

FINAL GRADE = .75 (QUIZ AVG.) + .25 (FINAL EXAM).

THE LIMITS OF THE OUTER SCALES ARE GIVEN BELOW. LOCATE AND CALIBRATE THE INTERIOR SCALE. SHOW A KEY TO DESCRIBE HOW THE NOMOGRAM IS USED. USE 10 SCALE.

1. WHAT WOULD BE A FINAL GRADE FOR A STUDENT WITH A 70 QUIZ AVERAGE AND A FINAL EXAM GRADE OF 90?
2. IF YOUR QUIZ AVERAGE WAS 50, HOW HIGH MUST YOU MAKE ON THE FINAL TO EARN A FINAL GRADE OF 60?
3. IF YOU HAD A QUIZ AVERAGE OF 90, HOW LOW COULD YOU SCORE ON THE EXAM AND EARN A FINAL GRADE OF 80?

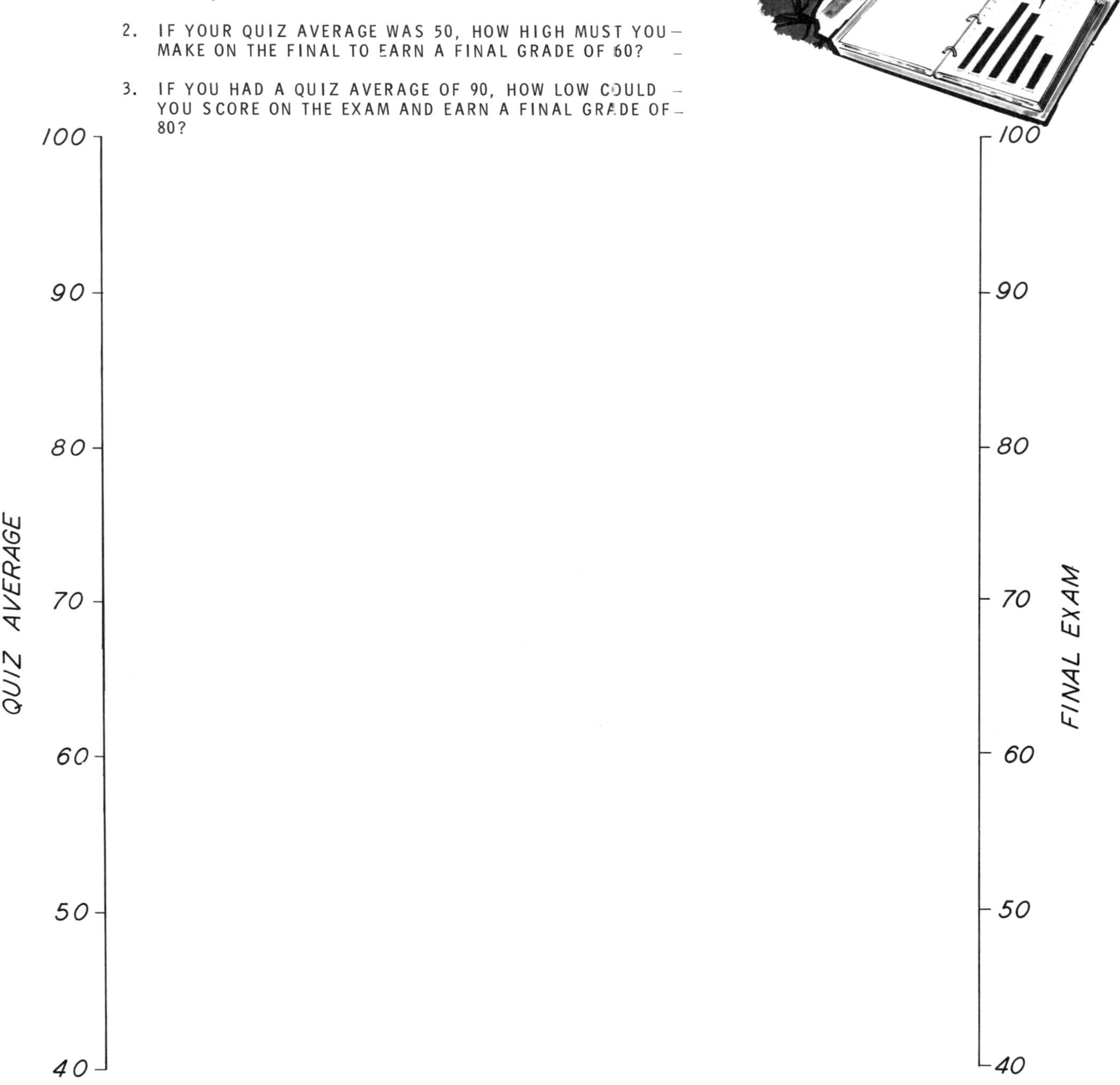

# LETTERING GUIDE LINES

THESE GUIDELINES CAN BE USED TO UNDERLAY TRACING PAPER OR OTHER TRANSLUSCENT PAPERS WHEN LETTERING A DRAWING.

| Graphics for Engineers © | NAME<br>FILE SEC DATE | MIN. | GRADE | |
|---|---|---|---|---|

## NOMOGRAM

NOMOGRAMS ARE USEFUL FOR RAPID CALCULATIONS THAT MUST BE MADE ON A REPETITIVE BASIS. SUCH A NOMOGRAM IS THE N-CHART THAT CAN BE CONSTRUCTED FOR EASILY COMPUTING MILES PER GALLON AFTER EACH FILL-UP OF GASOLINE. THIS RELATIONSHIP IS:

MILES PER GALLON = MILES ÷ GALLONS.

1. COMPLETE THE N-CHART BELOW AND CALIBRATE THE DIAGONAL SCALE.
2. PREPARE A KEY THAT DESCRIBES HOW THE CHART SHOULD BE USED.
3. ANSWER THE QUESTIONS BELOW.

EARLE


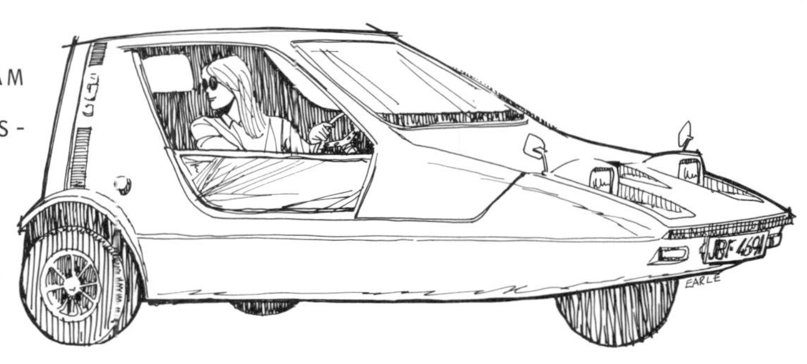

A. WHAT IS THE MPG FOR 275 MILES AND 19.2 GALLONS?

MPG = ___

B. IF YOU AVERAGE 15 MPG, HOW MANY GALLONS WOULD BE REQUIRED TO TRAVEL 225 MILES?

GALLONS = ___

C. IF YOU AVERAGE 12 MPG, HOW FAR WILL 15 GALLONS CARRY YOU?

MILES = ___

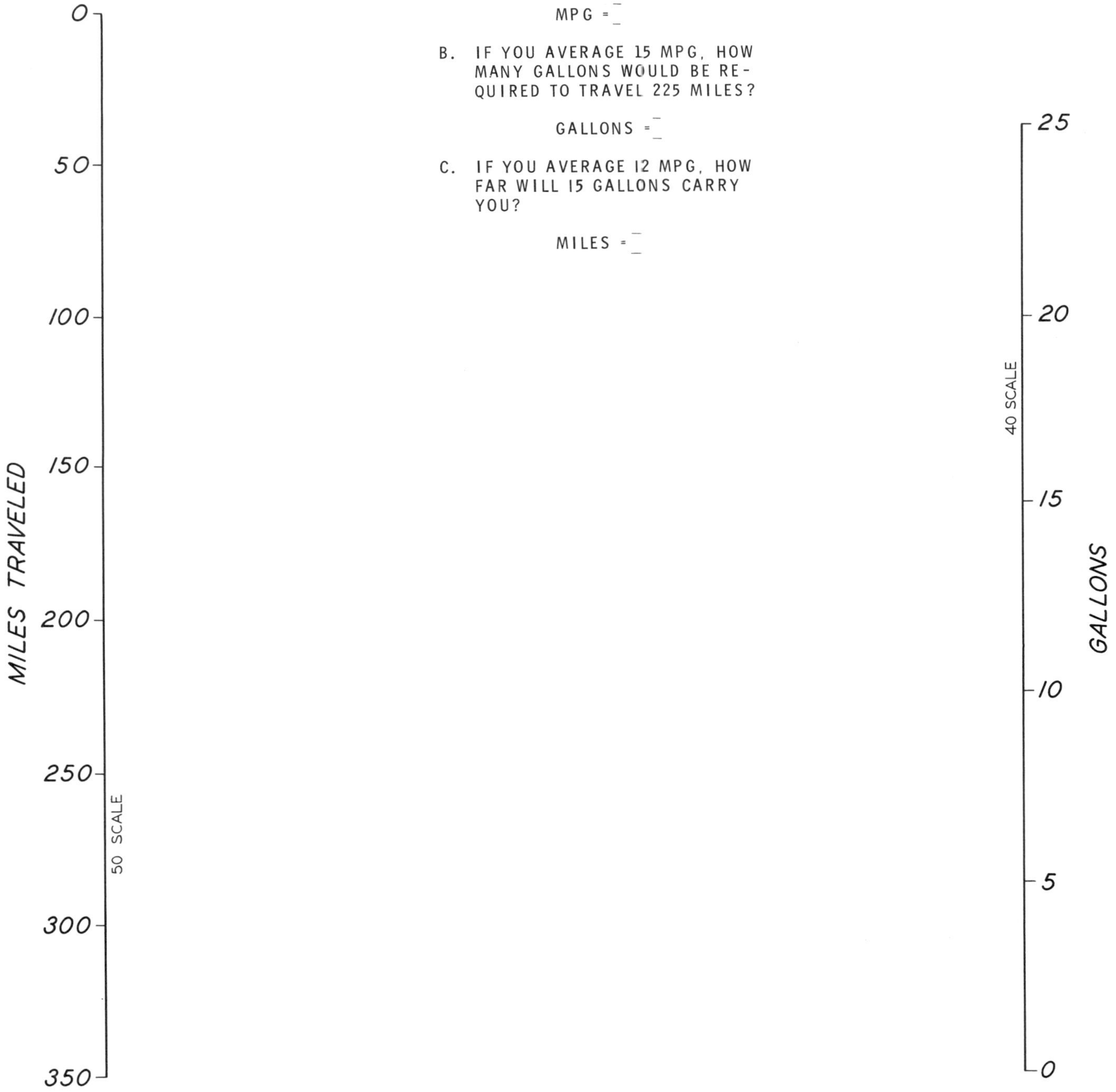

MILES TRAVELED
0
50
100
150
200
250
300
350
50 SCALE
GALLONS
25
20
15
10
5
0
40 SCALE


| Graphics & Geometry | NAME | MIN. | GRADE | |
|---|---|---|---|---|
| © | FILE SEC DATE | | | |

# LETTERING GUIDE LINES

THESE GUIDELINES CAN BE USED TO UNDERLAY TRACING PAPER OR OTHER TRANSLUSCENT PAPERS WHEN LETTERING A DRAWING.

| Graphics for Engineers © | NAME<br>FILE SEC DATE | MIN. | GRADE |
| --- | --- | --- | --- |

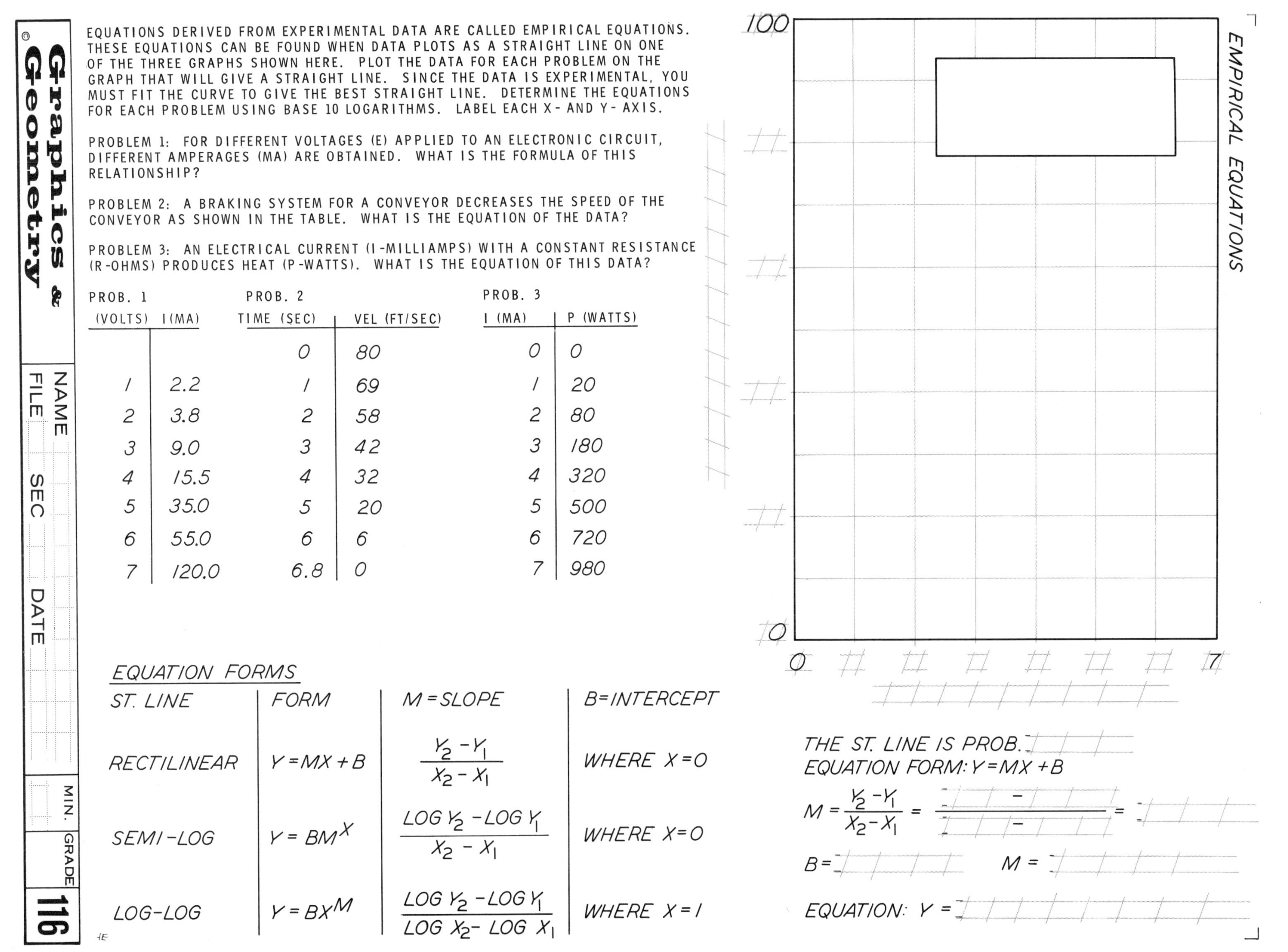

## EMPIRICAL EQUATIONS

EQUATIONS DERIVED FROM EXPERIMENTAL DATA ARE CALLED EMPIRICAL EQUATIONS. THESE EQUATIONS CAN BE FOUND WHEN DATA PLOTS AS A STRAIGHT LINE ON ONE OF THE THREE GRAPHS SHOWN HERE. PLOT THE DATA FOR EACH PROBLEM ON THE GRAPH THAT WILL GIVE A STRAIGHT LINE. SINCE THE DATA IS EXPERIMENTAL, YOU MUST FIT THE CURVE TO GIVE THE BEST STRAIGHT LINE. DETERMINE THE EQUATIONS FOR EACH PROBLEM USING BASE 10 LOGARITHMS. LABEL EACH X- AND Y- AXIS.

PROBLEM 1: FOR DIFFERENT VOLTAGES (E) APPLIED TO AN ELECTRONIC CIRCUIT, DIFFERENT AMPERAGES (MA) ARE OBTAINED. WHAT IS THE FORMULA OF THIS RELATIONSHIP?

PROBLEM 2: A BRAKING SYSTEM FOR A CONVEYOR DECREASES THE SPEED OF THE CONVEYOR AS SHOWN IN THE TABLE. WHAT IS THE EQUATION OF THE DATA?

PROBLEM 3: AN ELECTRICAL CURRENT (I-MILLIAMPS) WITH A CONSTANT RESISTANCE (R-OHMS) PRODUCES HEAT (P-WATTS). WHAT IS THE EQUATION OF THIS DATA?

PROB. 1

| (VOLTS) | I(MA) |
|---|---|
| 1 | 2.2 |
| 2 | 3.8 |
| 3 | 9.0 |
| 4 | 15.5 |
| 5 | 35.0 |
| 6 | 55.0 |
| 7 | 120.0 |

PROB. 2

| TIME (SEC) | VEL (FT/SEC) |
|---|---|
| 0 | 80 |
| 1 | 69 |
| 2 | 58 |
| 3 | 42 |
| 4 | 32 |
| 5 | 20 |
| 6 | 6 |
| 6.8 | 0 |

PROB. 3

| I (MA) | P (WATTS) |
|---|---|
| 0 | 0 |
| 1 | 20 |
| 2 | 80 |
| 3 | 180 |
| 4 | 320 |
| 5 | 500 |
| 6 | 720 |
| 7 | 980 |

### EQUATION FORMS

| ST. LINE | FORM | M = SLOPE | B = INTERCEPT |
|---|---|---|---|
| RECTILINEAR | $Y = MX + B$ | $\frac{Y_2 - Y_1}{X_2 - X_1}$ | WHERE $X = 0$ |
| SEMI-LOG | $Y = BM^X$ | $\frac{LOG\ Y_2 - LOG\ Y_1}{X_2 - X_1}$ | WHERE $X = 0$ |
| LOG-LOG | $Y = BX^M$ | $\frac{LOG\ Y_2 - LOG\ Y_1}{LOG\ X_2 - LOG\ X_1}$ | WHERE $X = 1$ |

100

0

0 7

THE ST. LINE IS PROB. ___

EQUATION FORM: $Y = MX + B$

$M = \frac{Y_2 - Y_1}{X_2 - X_1} = \frac{_ - _}{_ - _} = _$

$B = _$ $M = _$

EQUATION: $Y = _$

Graphics & Geometry ©

NAME  FILE  SEC  DATE  MIN.  GRADE

116

# LETTERING GUIDE LINES

THESE GUIDELINES CAN BE USED TO UNDERLAY TRACING PAPER OR OTHER TRANSLUSCENT PAPERS WHEN LETTERING A DRAWING.

| **Graphics for Engineers** © | NAME<br>FILE SEC DATE | MIN. | GRADE | |
|---|---|---|---|---|

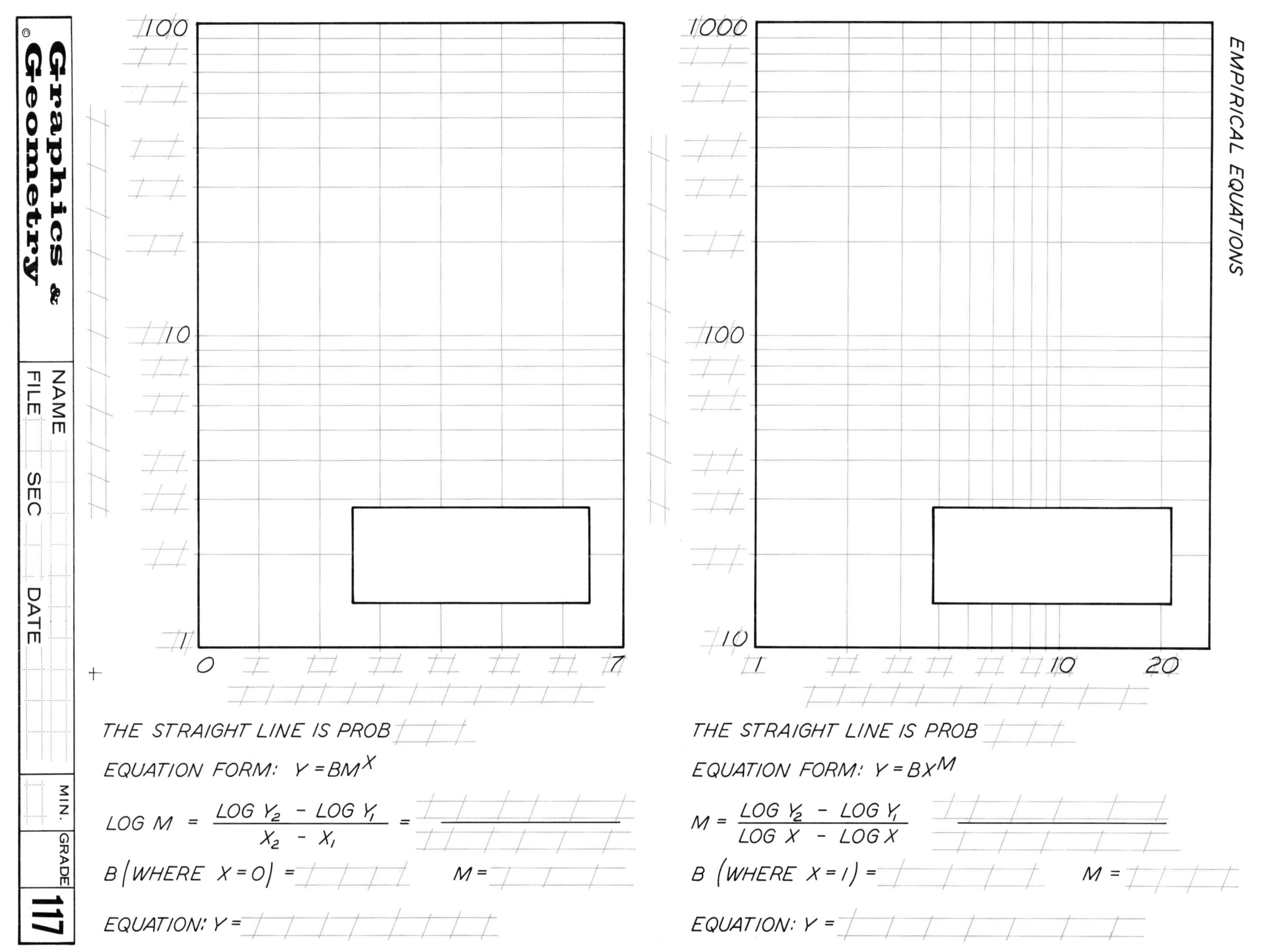


| Graphics & Geometry © | NAME | | MIN. | GRADE | 117 |
|---|---|---|---|---|---|
| | FILE | SEC DATE | | | |

# METRIC SCALES

Remove this page and fold it longwise along the desired scale for making metric measurements.

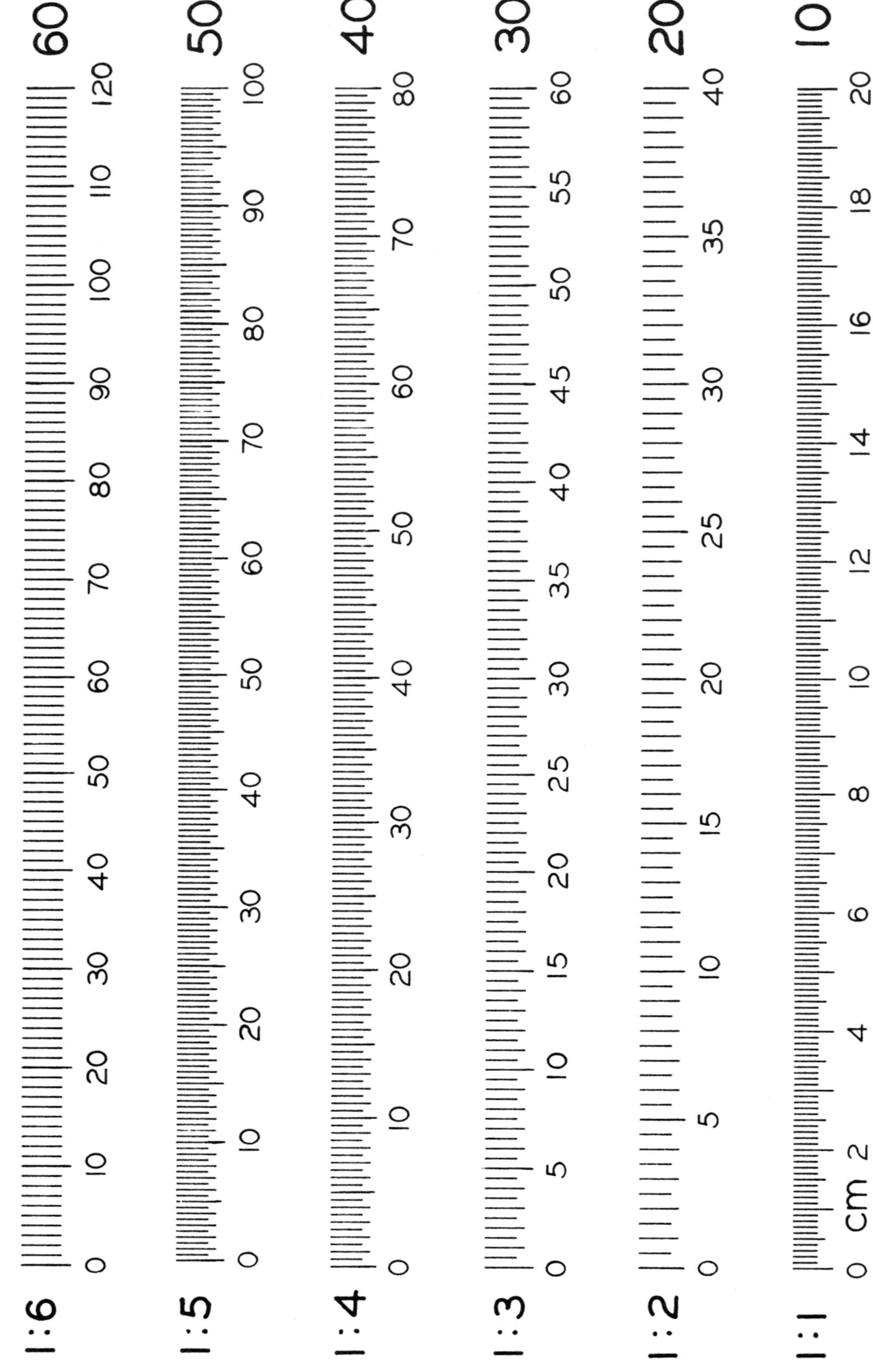

## GRAPHICAL DIFFERENTIATION

AN ASSEMBLY LINE SHUTTLE IS USED TO MOVE AUTOMOTIVE PARTS. IN DESIGNING THIS MECHANISM, THE ENGINEER MUST DETERMINE THE MAXIMUM ACCELERATION IN ORDER FOR THE FORCE REQUIRED TO BE FOUND, AND FINALLY THE TORQUE REQUIRED TO POWER THE SHUTTLE. THE VELOCITY IS PLOTTED BELOW FROM WHICH THE ACCELERATION CURVE MAY BE FOUND.

1. PLOT THE ACCELERATION CURVE BY GRAPHICAL DIFFERENTIATION.

2. WHAT IS THE MAXIMUM ACCELERATION? ____

3. AT WHAT TIME IS VELOCITY GREATEST? ____

   ACCELERATION LEAST? ____

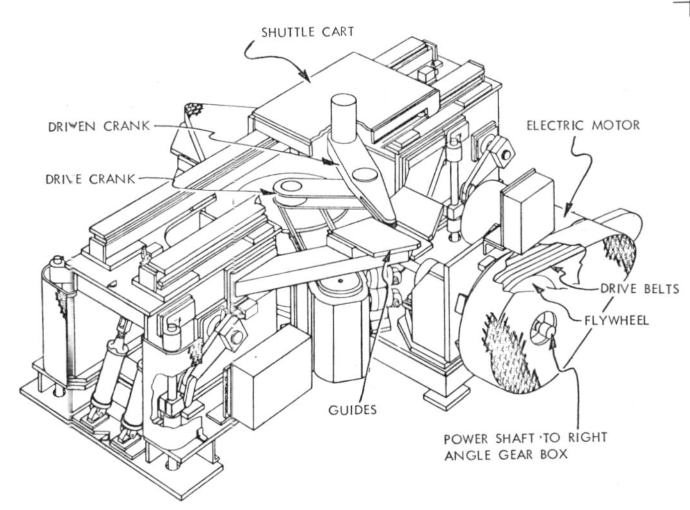

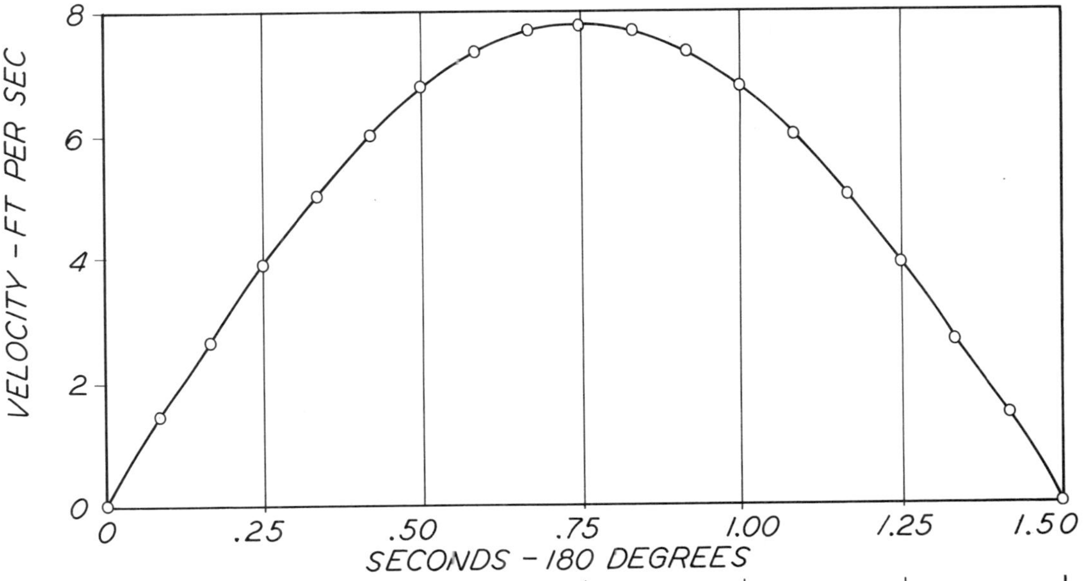

VELOCITY – FT PER SEC

8 6 4 2 0

0 .25 .50 .75 1.00 1.25 1.50

SECONDS – 180 DEGREES

15 10 5 0 −5 −10 −15

# METRIC SCALES

Remove this page and fold it longwise along the desired scale for making metric measurements.

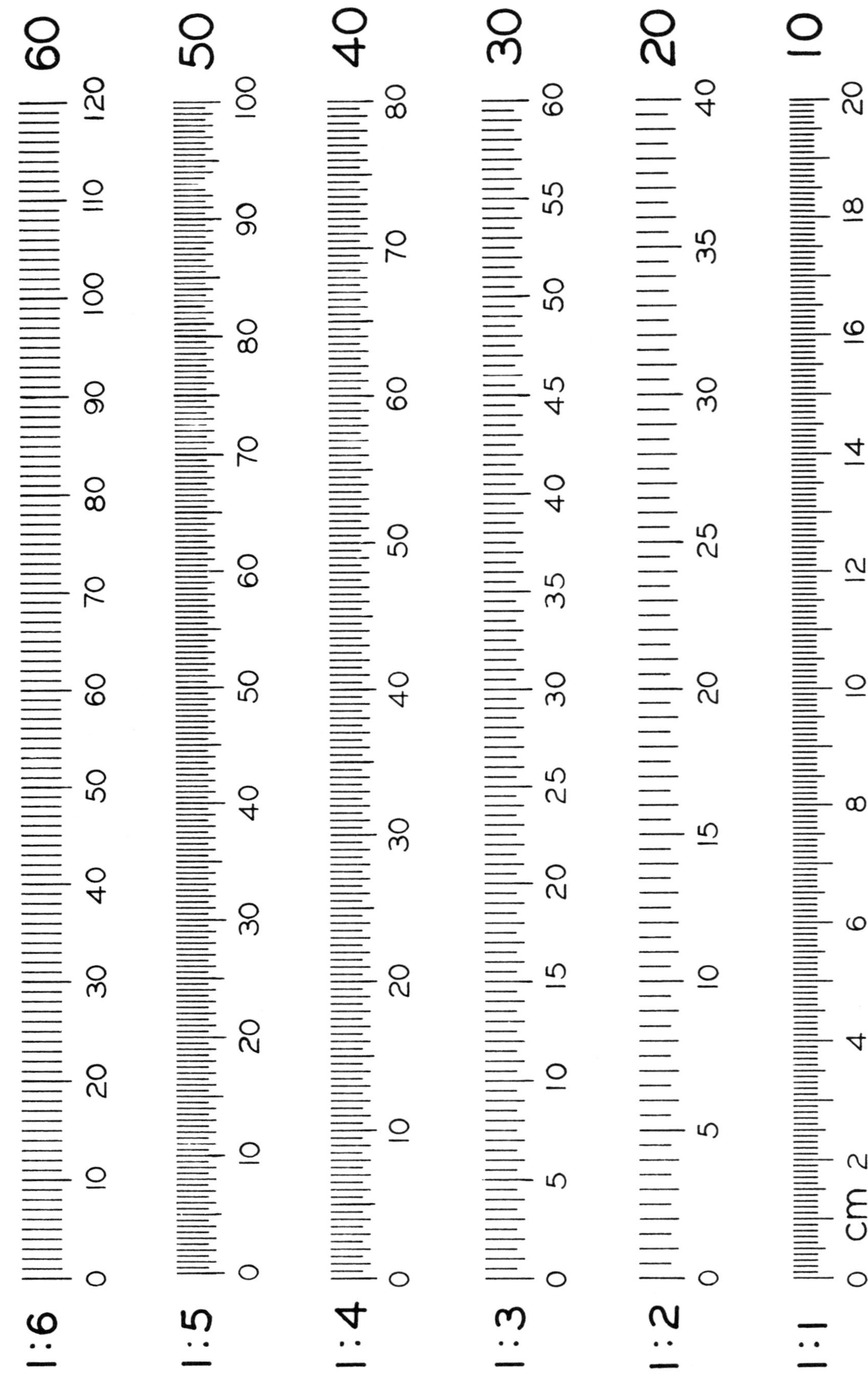

# GRAPHICAL INTEGRATION

A PLOT OF GROUND IS SHOWN IN THE PLAN BELOW. THE OWNER DESIRES TO ESTABLISH PROPERTY LINES PERPENDICULAR TO THE BASE LINE THAT WILL DIVIDE THE PLOT INTO THREE EQUAL AREA LOTS. PLOT THE INTEGRAL CURVE OF THIS PROPERTY AND LOCATE THE PROPERTY LINES. SHOW CONSTRUCTION.

1. WHAT IS THE AREA OF THE PROPERTY? —
2. WHAT IS THE AREA OF ONE LOT? —

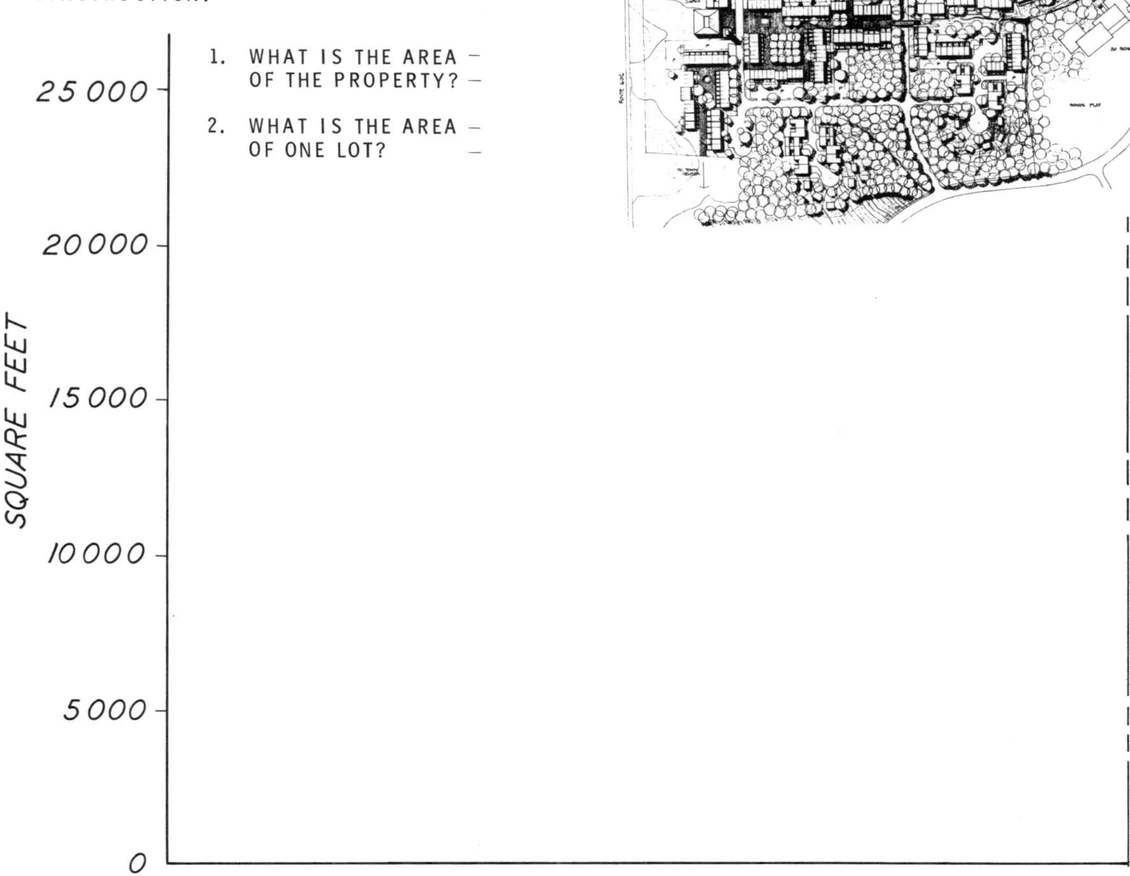


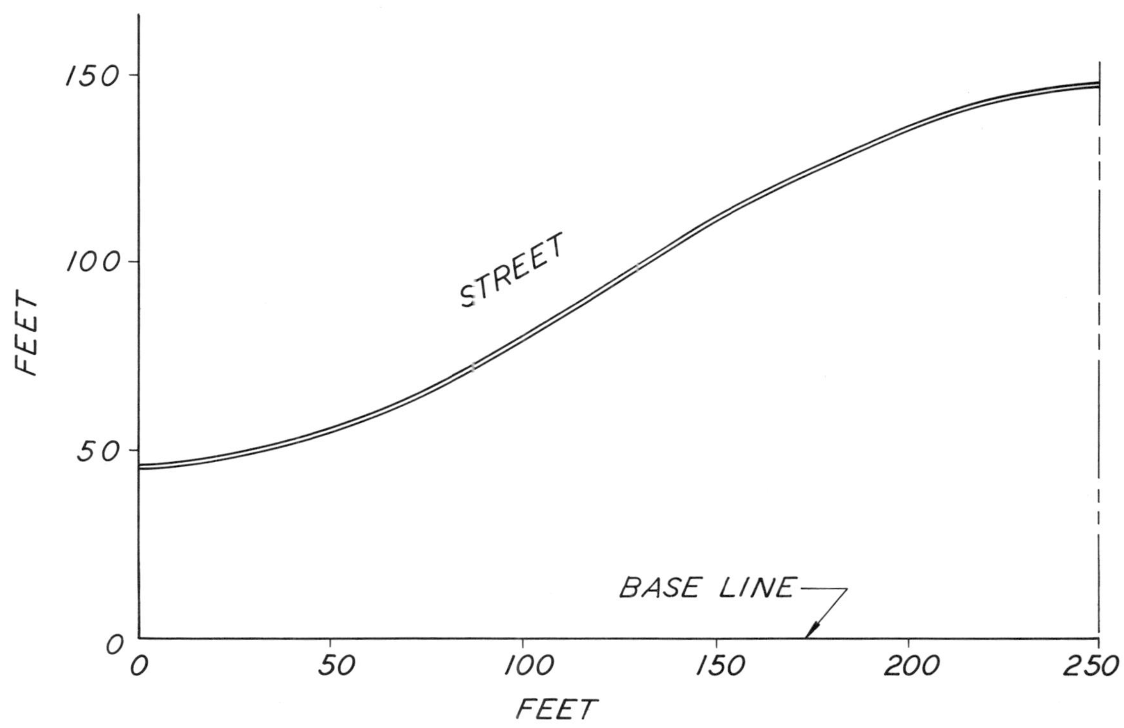


| **Graphics & Geometry** © | NAME | MIN. | GRADE | 119 |
|---|---|---|---|---|
| | FILE SEC DATE | | | |

# METRIC SCALES

Remove this page and fold it longwise along the desired scale for making metric measurements.

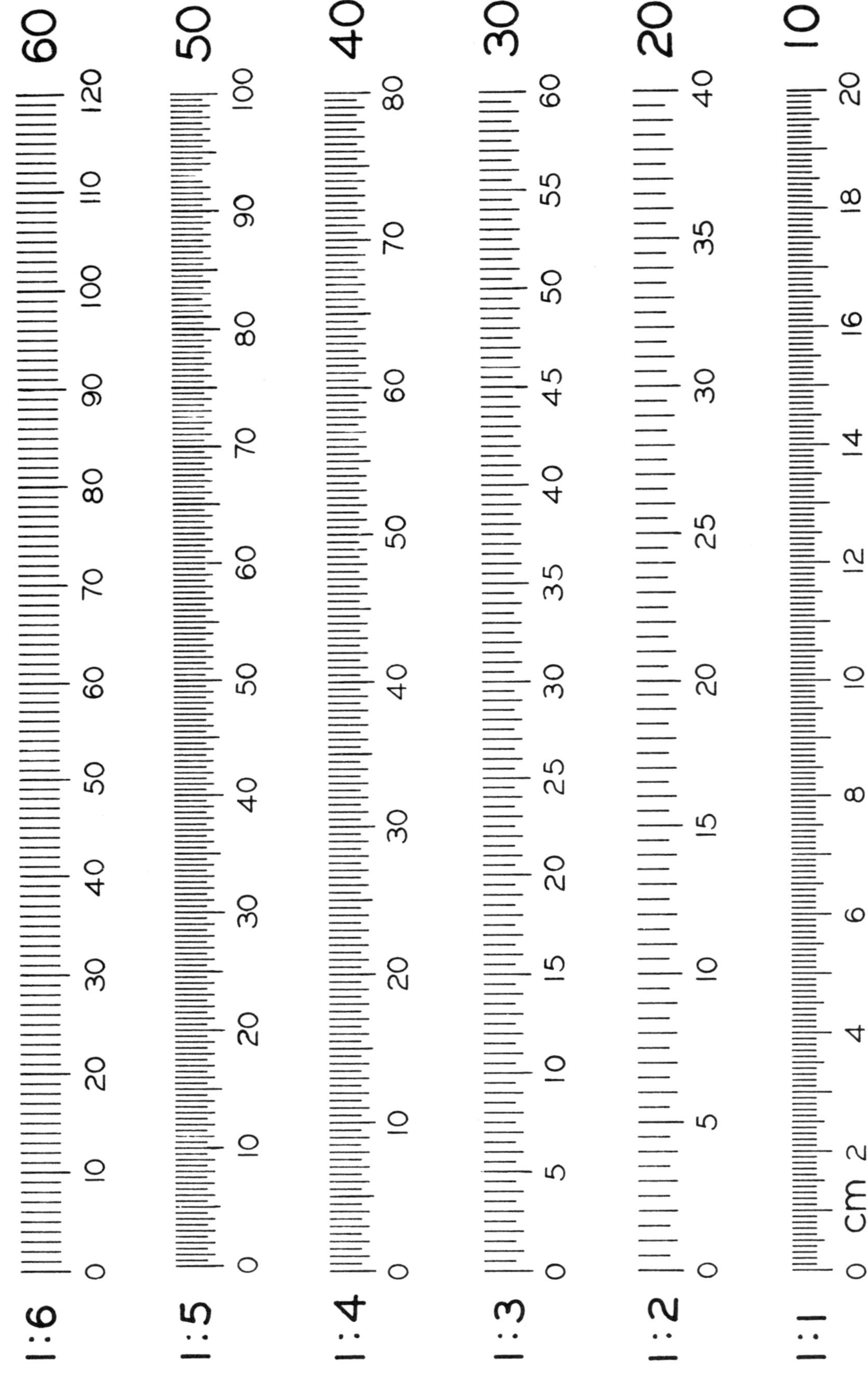

## DESIGN PROJECT

Title:

Course:

Section:

Date:

Team Members:

# PROBLEM IDENTIFICATION

1. Project title

2. Problem statement

3. Requirements and limitations

## 4. Needed information

DESIGN

## 5. Market Considerations

| Graphics for Engineers © | NAME<br>FILE SEC DATE | MIN. | GRADE | 2 |
|---|---|---|---|---|

DESIGN

# PRELIMINARY IDEAS

## 1. Brainstorming ideas

## 2. Description of best ideas

## 3. Attach sketches

1. Description of design

2. Attach scale drawings

ANALYSIS DESIGN

1. Function

2. Human engineering

3. Market & consumer acceptance

## 4. Physical description

## 5. Strength

## 6. Production procedures

## 7. Economic analysis

# DECISION

## 1. Decision table for evaluation

Design 1:

Design 2:

Design 3:

Design 4:

Design 5:

Design 6:

Design 7:

Design 8:

| Maximum value | Factors for analysis | Designs 1 | 2 | 3 | 4 | 5 | 6 | 7 | 8 |
|---|---|---|---|---|---|---|---|---|---|
| | Function | | | | | | | | |
| | Human factors | | | | | | | | |
| | Market analysis | | | | | | | | |
| | Strength | | | | | | | | |
| | Production procedures | | | | | | | | |
| | Cost | | | | | | | | |
| | Profitability | | | | | | | | |
| | Appearance | | | | | | | | |
| | | | | | | | | | |
| | | | | | | | | | |
| | | | | | | | | | |
| 10 | TOTALS | | | | | | | | |

DESIGN

# CONCLUSIONS

| Graphics for Engineers © | NAME<br>FILE SEC DATE | MIN. | GRADE | 10 |
|---|---|---|---|---|

| **Graphics for Engineers** © | NAME<br>FILE SEC DATE | MIN. | GRADE | |
|---|---|---|---|---|

| Graphics for Engineers © | NAME<br>FILE SEC DATE | MIN. | GRADE | |
|---|---|---|---|---|

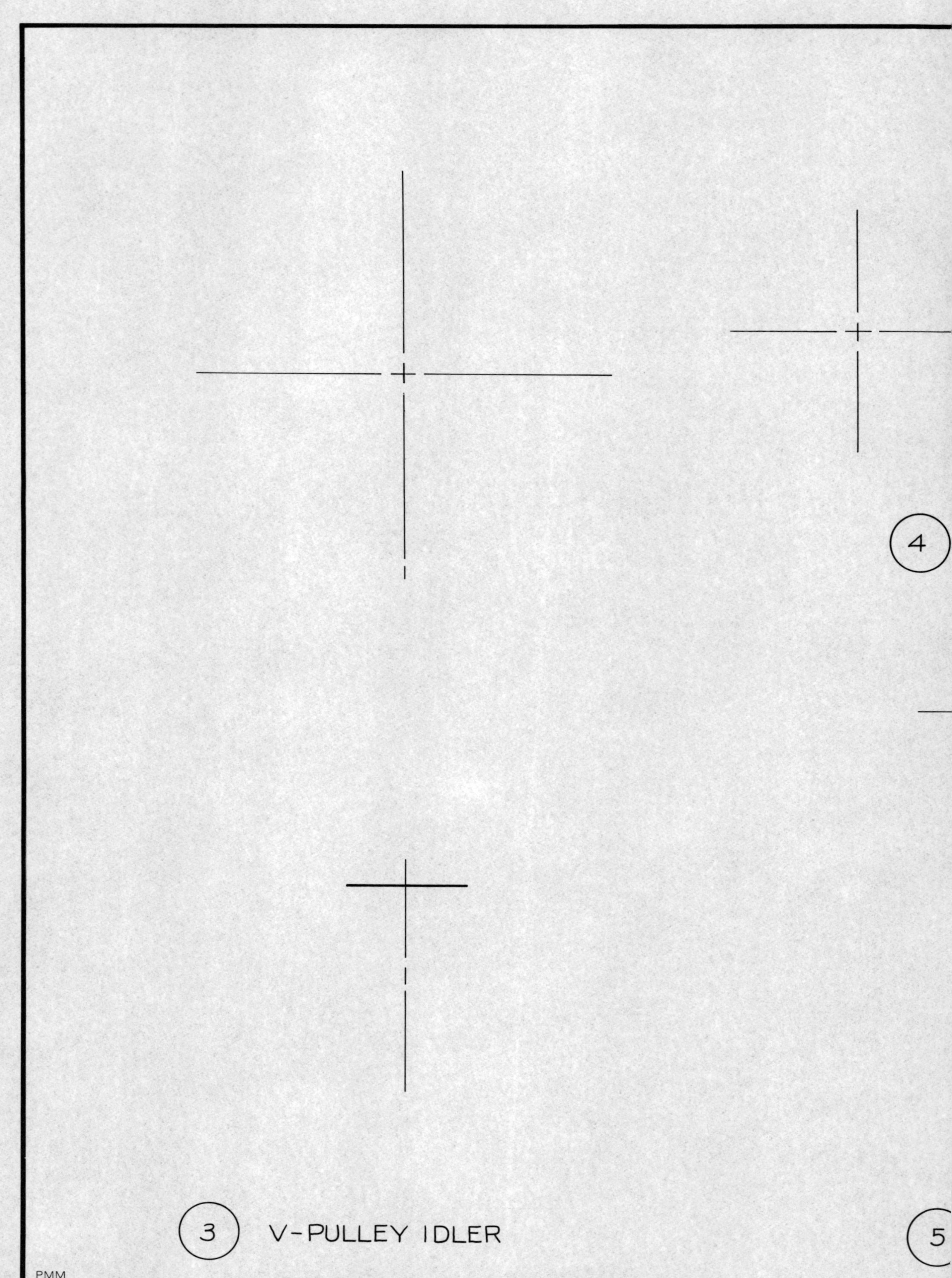
4
3
V-PULLEY IDLER
5
PMM

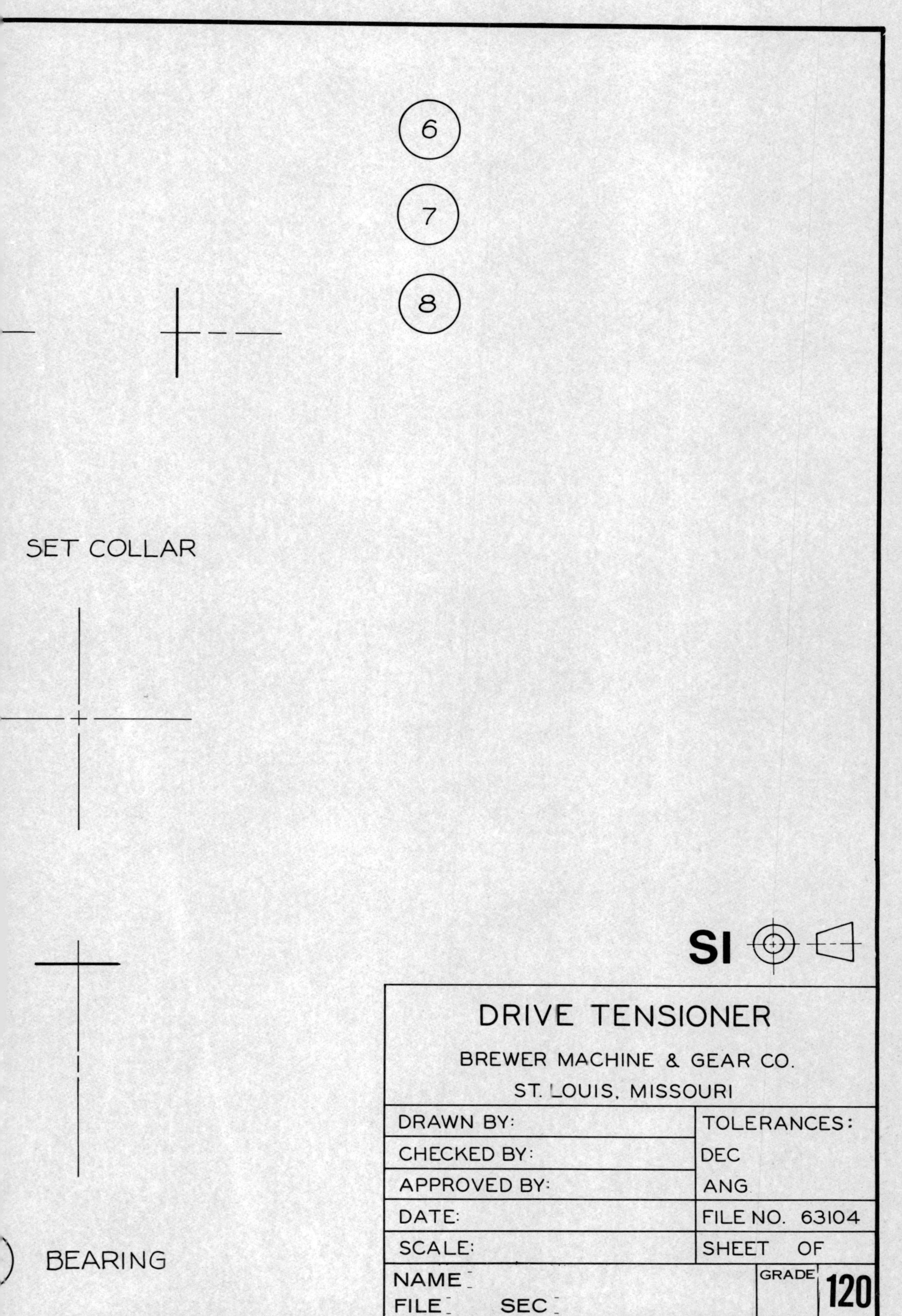
6
7
8
SET COLLAR
BEARING
SI
DRIVE TENSIONER
BREWER MACHINE & GEAR CO.
ST. LOUIS, MISSOURI
DRAWN BY:
CHECKED BY:
APPROVED BY:
DATE:
SCALE:
TOLERANCES:
DEC
ANG
FILE NO. 63104
SHEET OF
NAME
FILE
SEC
GRADE
120

6

7

8

SET COLLAR

BEARING

SI

DRIVE TENSIONER

BREWER MACHINE & GEAR CO

ST. LOUIS, MISSOURI

| DRAWN BY: | TOLERANCES: |
|---|---|
| CHECKED BY: | DEC |
| APPROVED BY: | ANG |
| DATE: | FILE NO. 63104 |
| SCALE: | SHEET OF |
| NAME: FILE: SEC: | GRADE 120 |

IDLER BASE

PMM

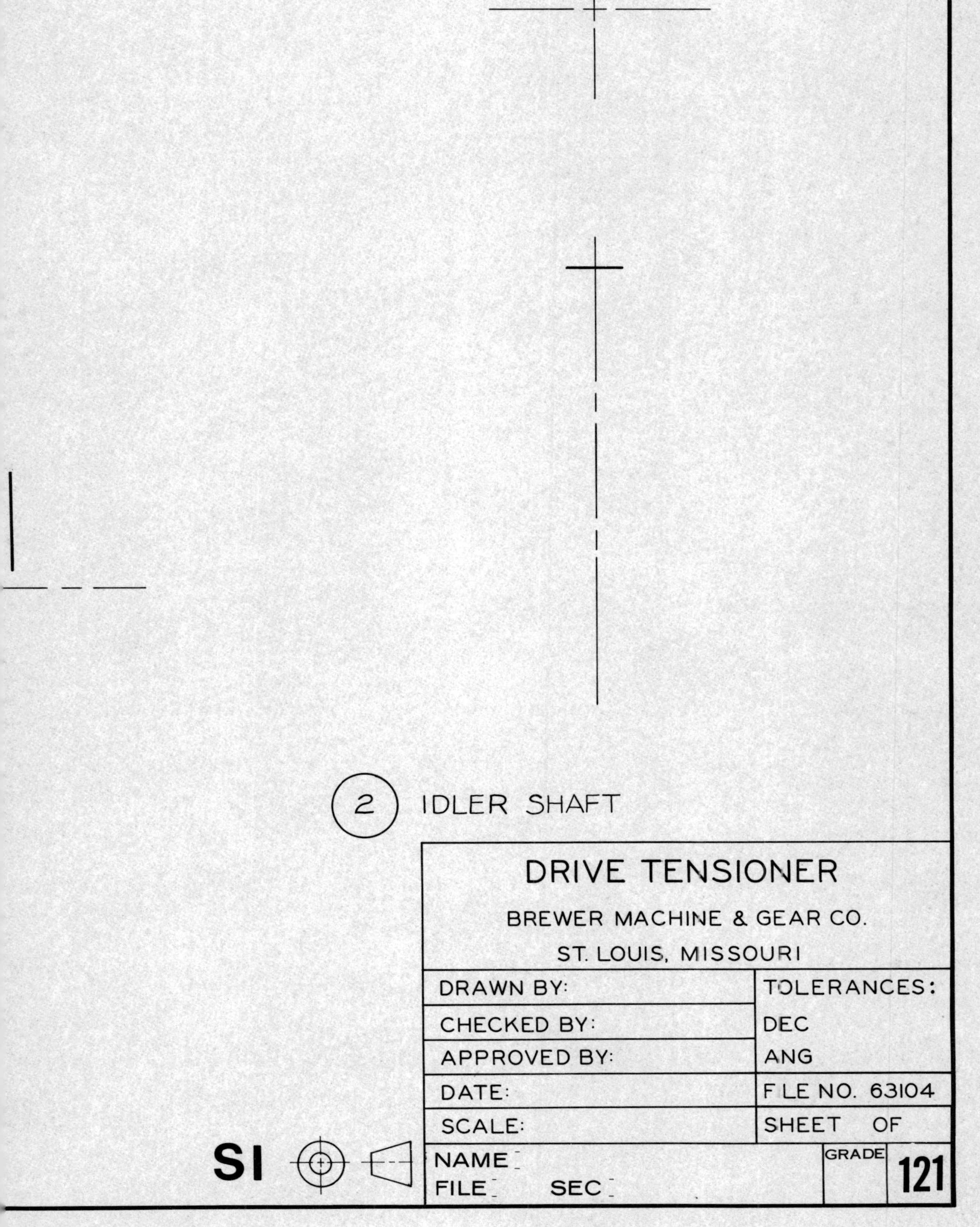

2 IDLER SHAFT
DRIVE TENSIONER
BREWER MACHINE & GEAR CO.
ST. LOUIS, MISSOURI
DRAWN BY:
CHECKED BY:
APPROVED BY:
DATE:
SCALE:
TOLERANCES:
DEC
ANG
FILE NO. 63104
SHEET OF
SI
NAME
FILE
SEC
GRADE
121

(2) IDLER SHAFT

DRIVE TENSIONER

BREWER MACHINE & GEAR CO.

ST LOUIS MISSOURI

| | |
|---|---|
| DRAWN BY | TOLERANCES: |
| CHECKED BY: | DEC |
| APPROVED BY | ANG |
| DATE: | FILE NO. 63104 |
| SCALE: | SHEET OF |
| NAME<br>FILE SEC | GRADE 121 |

S1

| NO | PART | REQ | MATL |
|---|---|---|---|
| | | | |

PARTS LIST

# DRIVE TENSIONER

BREWER MACHINE & GEAR CO.

ST. LOUIS, MISSOURI

| DRAWN BY: | TOLERANCES: |
|---|---|
| CHECKED BY: | DEC |
| APPROVED BY: | ANG |
| DATE: | FILE NO. 63104 |
| SCALE: | SHEET OF |

NAME

FILE SEC

GRADE 122